Vitamin K - Recent Topics on the Biology and Chemistry

Edited by Hiroyuki Kagechika and Hitoshi Shirakawa

Published in London, United Kingdom

IntechOpen

Supporting open minds since 2005

Vitamin K - Recent Topics on the Biology and Chemistry
http://dx.doi.org/10.5772/intechopen.82971
Edited by Hiroyuki Kagechika and Hitoshi Shirakawa

Contributors
Ebtesam Abdullah Al-Suhaimi, Meneerah A. Aljafary, Hussah Alshwyeh, Nada Alahmadi, Adeeb Shehzad, Huseyin Tombuloglu, Zagit Gaymalov, Abdelgadir Homeida, Shashank S. Jadhav, Dilip Mehta, Anselm de Souza, Shotaro Goto, Shuichi Setoguchi, Kazuhisa Matsunaga, Jiro Takata, Nikolaos Schizas, Theodoros Michailidis, Asterios Karakanas, Petros Keryttopoulos, Shinya Fujii, Yoshitomo Suhara, Hiroyuki Kagechika, Satyanarayana Battula

First published in London, United Kingdom, 2022 by IntechOpen
IntechOpen is the global imprint of INTECHOPEN LIMITED, registered in England and Wales, registration number: 11086078, 5 Princes Gate Court, London, SW7 2QJ, United Kingdom
Printed in Croatia

British Library Cataloguing-in-Publication Data
A catalogue record for this book is available from the British Library

Additional hard and PDF copies can be obtained from orders@intechopen.com

Vitamin K - Recent Topics on the Biology and Chemistry
Edited by Hiroyuki Kagechika and Hitoshi Shirakawa
p. cm.

This title is part of the Biochemistry Book Series, Volume 27
Topic: Metabolism
Series Editor: Miroslav Blumenberg
Topic Editors: Yannis Karamanos

Print ISBN 978-1-83969-391-5
Online ISBN 978-1-83969-392-2
eBook (PDF) ISBN 978-1-83969-393-9
ISSN 2632-0983

For EU product safety concerns:
IN TECH d.o.o., Prolaz Marije Krucifikse Kozulić 3, 51000 Rijeka, Croatia, info@intechopen.com or visit our website at intechopen.com.

IntechOpen Book Series
Biochemistry
Volume 27

Aims and Scope of the Series

Biochemistry, the study of chemical transformations occurring within living organisms, impacts all of the life sciences, from molecular crystallography and genetics, to ecology, medicine and population biology. Biochemistry studies macromolecules - proteins, nucleic acids, carbohydrates and lipids –their building blocks, structures, functions and interactions. Much of biochemistry is devoted to enzymes, proteins that catalyze chemical reactions, enzyme structures, mechanisms of action and their roles within cells. Biochemistry also studies small signaling molecules, coenzymes, inhibitors, vitamins and hormones, which play roles in the life process. Biochemical experimentation, besides coopting the methods of classical chemistry, e.g., chromatography, adopted new techniques, e.g., X-ray diffraction, electron microscopy, NMR, radioisotopes, and developed sophisticated microbial genetic tools, e.g., auxotroph mutants and their revertants, fermentation, etc. More recently, biochemistry embraced the 'big data' omics systems.
Initial biochemical studies have been exclusively analytic: dissecting, purifying and examining individual components of a biological system; in exemplary words of Efraim Racker, (1913 –1991) "Don't waste clean thinking on dirty enzymes." Today, however, biochemistry is becoming more agglomerative and comprehensive, setting out to integrate and describe fully a particular biological system. The 'big data' metabolomics can define the complement of small molecules, e.g., in a soil or biofilm sample; proteomics can distinguish all the proteins comprising e.g., serum; metagenomics can identify all the genes in a complex environment e.g., the bovine rumen. This Biochemistry Series will address both the current research on biomolecules, and the emerging trends with great promise.

Meet the Series Editor

Miroslav Blumenberg, Ph.D., was born in Subotica and received his BSc in Belgrade, Yugoslavia. He completed his Ph.D. at MIT in Organic Chemistry; he followed up his Ph.D. with two postdoctoral study periods at Stanford University. Since 1983, he has been a faculty member of the RO Perelman Department of Dermatology, NYU School of Medicine, where he is codirector of a training grant in cutaneous biology. Dr. Blumenberg's research is focused on the epidermis, expression of keratin genes, transcription profiling, keratinocyte differentiation, inflammatory diseases and cancers, and most recently the effects of the microbiome on the skin. He has published more than 100 peer-reviewed research articles and graduated numerous Ph.D. and postdoctoral students.

Meet the Volume Editors

Hiroyuki Kagechika received his bachelor's degree and Ph.D. in Pharmaceutical Sciences from the University of Tokyo, Japan, where he served as an associate professor until 2004. He is currently a professor at the Institute of Biomaterials and Bioengineering (IBB), Tokyo Medical and Dental University (TMDU). From 2010 to 2012, he was the dean of the Graduate School of Biomedical Science. Since 2012, he has served as the vice dean of the Graduate School of Medical and Dental Sciences. He has been the director of the IBB since 2020. Dr. Kagechika's major research interests are the medicinal chemistry of retinoids, vitamins D/K, and nuclear receptors. He has developed various compounds including a drug for acute promyelocytic leukemia.

Hitoshi Shirakawa is a professor at the Laboratory of Nutrition, Graduate School of Agricultural Science, Tohoku University, Japan. He obtained his Ph.D. from the same university in 1996 with his studies on the structure and molecular functions of chromosomal protein HMG2. His current research is focused on the modulation of intracellular signaling by nutrients and other food ingredients, especially on the alternation of inflammatory response and steroidogenesis by vitamin K, biotin, isoprenoids, and other food ingredients. Dr. Shirakawa also investigates the effects of fermented rice/wheat bran extract on the prevention of lifestyle-related diseases and identifies novel functional ingredients from fermented materials.

Contents

Preface

Vitamin K is a fat-soluble vitamin with a 1,4-naphthoquinone structure bearing various side-chain moieties that is responsible for blood coagulation and bone homeostasis. It acts as the cofactor of γ-glutamyl carboxylase, which catalyzes the γ-carboxylation of glutamate residue of vitamin K-dependent proteins, including factors II (prothrombin), VII, IX, and X. The proteins that have γ-carboxylated glutamate residue are also found in the bone matrix and aortic vessel wall and are involved in the regulation of these structures' calcification.

In the past two decades, novel biological activities of vitamin K other than the coenzyme function of γ-glutamyl carboxylase have been reported. For example, vitamin K shows an anti-inflammatory effect via the inhibition of the NF-κB pathway, and steroidogenesis by the activation of cAMP-dependent protein kinase. In the vitamin K family, menaquinone-4 (MK-4) plays important roles in many biological systems, including enhancement of testosterone production and inhibition of arteriosclerosis and tumor progression. Further, MK-4 was identified as the specific ligand for the steroid and xenobiotic receptor (SXR) and it modulates the transcription of genes in osteoblast, hepatocyte, and intestinal cells. These novel functions of vitamin K could prevent and control exacerbations in non-communicable diseases and might contribute to prolonged life expectancy.

Ingested vitamin K from plants, animals, and fermented products is endogenously converted into MK-4. UbiA prenyltransferase domain-containing protein 1 (UBIAD1) was identified as a key enzyme for biosynthesis of MK-4. An unknown factor cleaves the side chain of vitamin K and generates 2-methyl-1,4-naphthoquinone (vitamin K3, menadione). UBIAD1 catalyzes MK-4 production from menadione and geranyl-geranyl diphosphate derived from the mevalonate pathway. In organs of animals, including humans, MK-4 is the predominant form of vitamin K. These findings on non-classical activities of vitamin K indicate the existence of unrevealed mechanisms of action for physiological and pharmacological functions of vitamin K. In the case of vitamins A and D, the identification of active metabolites (retinoic acid for vitamin A, and 1α,25-dihydroxyvitamin D3 for vitamin D) and the target nuclear receptors caused dramatic improvement in elucidation of their hormonal functions and clinical applications, while the metabolic pathway of vitamin K and properties of metabolites, including the possible unidentified active metabolites, remains unclear,

Vitamin K is an attractive lead compound for drug discovery. Several vitamin K derivatives have been synthesized, including possible metabolites, most of them consisting of structural features like those of natural vitamin K. Development of novel vitamin K derivatives with unique chemical structure and biological profile, especially those that elicit non-classical vitamin K functions, would lead to novel clinical applications of vitamin K.

This book discusses the biology and chemistry of vitamin K, which is helpful for fundamental and clinical investigations. I would like to sincerely thank all the authors who contributed to this book.

Hitoshi Shirakawa, Ph.D.
Professor,
Graduate School of Agricultural Science,
Tohoku University,
Sendai, Japan

Hiroyuki Kagechika, Ph.D.
Professor,
Institute of Biomaterials and Bioengineering,
Tokyo Medical and Dental University,
Tokyo, Japan

Chapter 1

Recent Advances in the Medicinal Chemistry of Vitamin K Derivatives: An Overview (2000–2021)

Shinya Fujii, Yoshitomo Suhara and Hiroyuki Kagechika

Abstract

In recent decades, many physiological and pharmacological functions of vitamin K other than its role as the cofactor of γ-glutamyl carboxylase (GGCX) have been identified, and consequently, many vitamin K derivatives and related congeners, including putative metabolites, have been designed and synthesized. Their biological activities include antitumor activity, anti-inflammatory activity, neuroprotective effects, neural differentiation-inducing activity, and modulating potency toward the nuclear steroid and xenobiotic receptor (SXR). These activities make vitamin K and its derivatives attractive candidates for drug discovery. In this chapter, an overview of recent advances in the medicinal chemistry of vitamin K, focusing especially on SXR modulation, neural differentiation, and antitumor activities, was provided.

Keywords: metabolites, synthetic analogs; neural differentiation, nuclear receptor, steroid and xenobiotic receptor, antitumor, phthalazine-1,4-dione

1. Introduction

Vitamin K is the term used to describe derivatives of naphthoquinones **1–4** [1, 2]. It was originally identified as a specific cofactor for γ-glutamyl carboxylase (GGCX), which catalyzes the formation of γ-carboxyglutamyl (Gla) residues in vitamin K-dependent proteins. Since then, various other biological activities of vitamin K have been reported. For example, antitumor activity of vitamin K_3 (**4**: menadione: 2-methyl-1,4-naphthoquinone) was reported in the 1980s, [3–6] and antitumor activity of vitamin K_2 (**2**: menaquinone-n; MK-n) was also found in the 1990s [7, 8]. Among the homologs of vitamin K, menaquinone-4 (**3**, MK-4), which contains four isoprene units in the side chain, has been most intensively investigated [9–11]. MK-4 binds to human pregnane X nuclear receptor (PXR), which is also called the steroid and xenobiotic receptor (SXR), and regulates transcription of osteoblastic genes [12, 13]. MK-4 and its derivatives also have roles in neural differentiation, as well as neuroprotective effects [14]. In addition, MK-4 exhibits anti-inflammatory activity by suppressing the NF-κB pathway [15], exerts an inhibitory effect on arteriosclerosis [16], and shows growth-inhibitory activity toward hepatocellular carcinoma (HCC) cells [17, 18]. Thus, vitamin K and its derivatives are attractive candidates

 IntechOpen

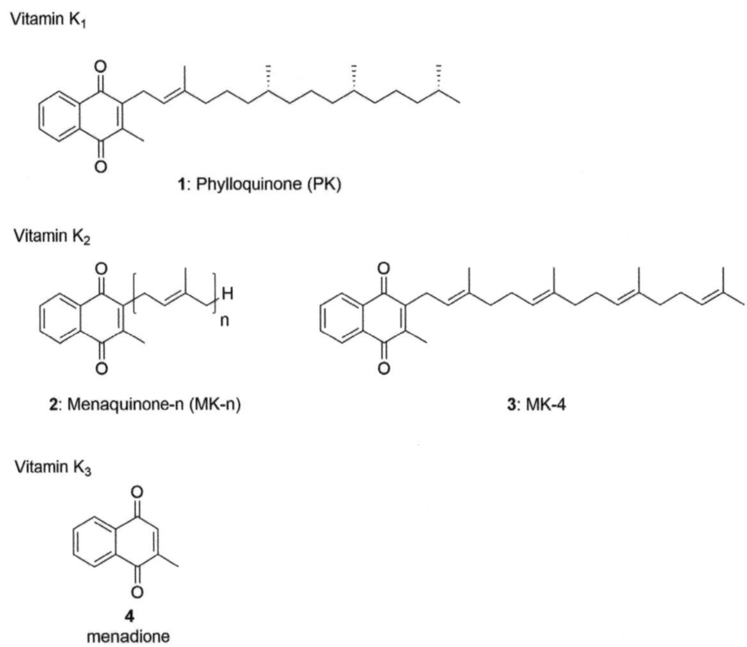

Figure 1.
Structures of vitamin K homologs.

for drug discovery. The following sections provide a review and perspective of the medicinal chemistry of vitamin K derivatives mainly developed from the beginning of this century (**Figure 1**).

2. Structure–activity relationship of vitamin K analogs for transcriptional activity through nuclear receptor SXR

There are two kinds of natural vitamin K homologs, phylloquinone (PK) (**1**) and menaquinones (MK-n) (**2**). Since the discovery of vitamin K, research has mainly focused on its role in the blood coagulation system. However, it was recently revealed that MK-4 (**3**) binds to the steroid and xenobiotic receptor (SXR, pregnane X receptor in mice: PXR), a member of the nuclear receptor superfamily, and exhibits agonist activity [19]. The mechanism has been reported to be as follows; first, **3** binds to SXR and forms a heterodimer with retinoid X receptor (RXR). Then, the heterodimer binds to SXR-responsive elements on DNA and gene expression of a drug-metabolizing enzyme, CYP3A4, is induced [12]. The menaquinone **3** also induces gene expression of proteins involved in osteogenesis through binding to SXR [13]. Other menaquinones showed similar effects, but **3** was the most potent. Interestingly, **1** did not have such an effect.

Following describes several vitamin K derivatives synthesized for structure–activity relationship studies of SXR agonists. Focusing on the double bonds and methyl groups in the side chain of MK-4, compounds **5–14**, in which the double bonds in the isoprene side chain were progressively saturated or methyl groups were deleted, were synthesized (**Figure 2**) [20]. In this study, the deuterium-labeled compounds were employed since these compounds enable us to investigate the conversion rate of the analogs to MK-4. The SXR-mediated transcriptional activity of each compound was evaluated in two ways, using SXR-GAL4 and CYP3A4 promoters. The results showed that as the number of double bonds was decreased,

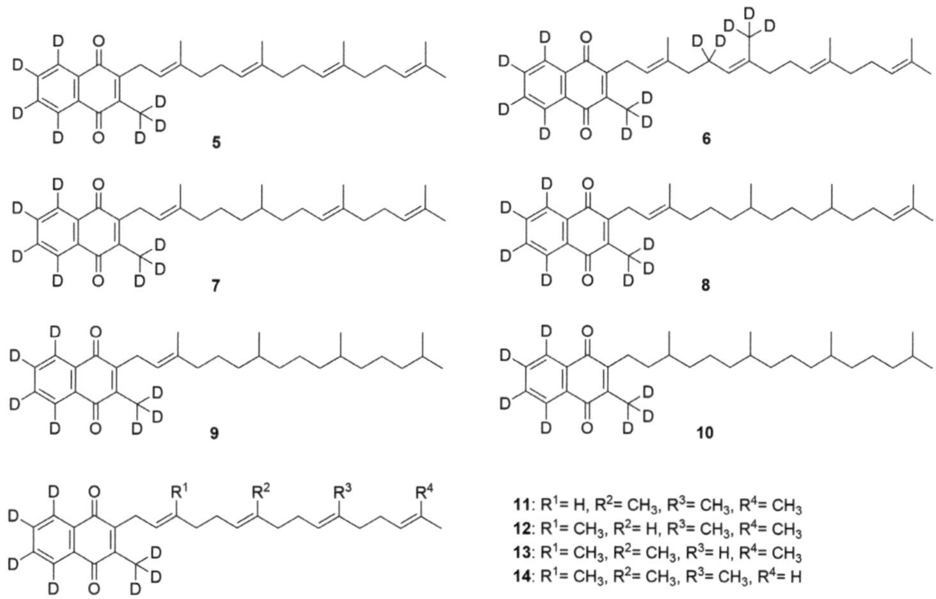

Figure 2.
Structure of vitamin K analogs **5–14**.

the transcriptional activity also significantly decreased [20]. This tendency was particularly pronounced with the CYP3A4 promoter assay. Deletion of methyl groups on the side chain also decreased the transcriptional activity. These results indicate that both the methyl groups and double bonds of the side chain of MK-4 are important for SXR-mediated transcriptional activity (**Figure 2**).

Since the isoprene structure of the side chain of menaquinones is important for the activity, vitamin K derivatives **15–19**, in which an isoprene side chain is symmetrically introduced into the naphthoquinone part, were next synthesized (**Figure 3**) [21]. The transcriptional activity of these analogs peaked at compound **17**, which contains two side chains of MK-2, and then remarkably decreased with increasing length of the side chains [20–22]. These results indicate that the transcriptional activity of vitamin K derivatives is greatly affected by the length and bulk of the side chains (**Figure 3**).

Then, in order to investigate how the transcriptional activity changes depending on the polarity of the side chain, vitamin K analogs introduced hydrophilic or

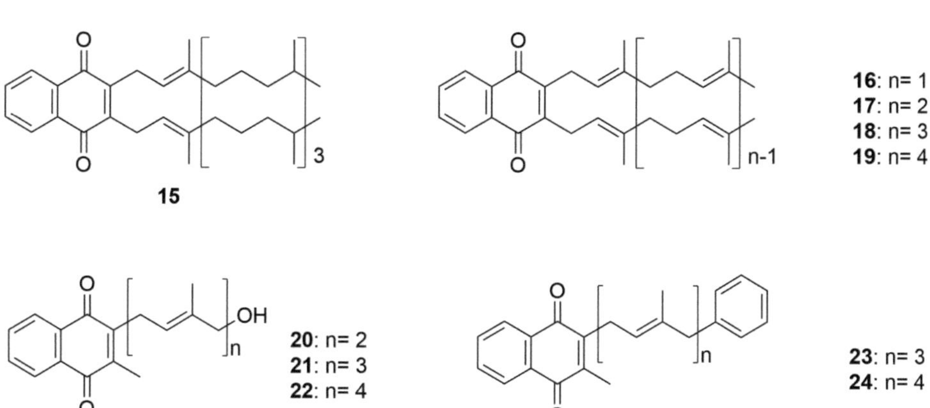

Figure 3.
Vitamin K analogs 15–24.

hydrophobic functional groups at the end of the side chain; namely, compounds **20–22** with a hydroxyl group as a hydrophilic functional group and compounds **23** and **24** with a phenyl group as a hydrophobic functional group, were synthesized (**Figure 2**) [23]. The transcriptional activity of compounds **20–22** was decreased; on the other hand, that of compounds **23** and **24** was markedly increased. In particular, compound **23**, an analog of MK-3 bearing a phenyl group at the end of the side chain, showed comparable activity to that of rifampicin, a known SXR ligand. Computational analysis of the binding states of the vitamin K derivative **23** with the ligand-binding site of SXR using the MOE (Molecular Operating Environment)-integrated computational chemistry system indicated that the oxygen atoms of the quinone part of **23** form hydrogen bonds with His407 and Ser247 of SXR [23]. In view of their potent activity, MK-3 and MK-4 were expected to show similar interactions. Thus, the SXR-mediated transcriptional activity of vitamin K requires an appropriate length and bulk of the isoprene side chain, and the side chain structure also has a significant influence [20, 21, 23].

3. Neuronal differentiation-inducing activity of vitamin K and its analogs

MK-4 is present at relatively high concentrations in the brain, though its physiological role remains unclear. As one of the biological action in the brain, it has been reported that it protects neurons against oxidative stress [14, 24–26]. It is also known that neural stem cells differentiate into neuronal progenitors and glial progenitors, and then, neuronal progenitors differentiate into neurons, while glial progenitors differentiate into astrocytes and oligodendrocytes [27]. Recently, it has been found that menaquinones selectively induce the differentiation of neural progenitors into neurons, although their potency was not high [28]. This activity differed depending on the repeat structure of the isoprene side chain of the menaquinones. Therefore, if this activity can be increased by derivatization of vitamin K, it might be possible to regulate differentiation using safe and small molecule inducers of neural differentiation. Thus, new vitamin K derivatives that would induce differentiation of neural stem cells into neurons were explored.

Considering the lipophilic environment of the brain, vitamin K analogs bearing various hydrophobic functional groups such as benzene or naphthalene in the side chain were designed and synthesized (**Figure 4**). The compounds were evaluated for neuronal differentiation-inducing activity toward stem cells derived from mouse fetal cerebrum. After the compounds were added to the cells and the cells were cultured, the expression levels of Map2 and Gfap, which are expressed specifically in neurons and astrocytes, were quantified by real-time PCR. Interestingly, most synthesized compounds showed a significant increase in the induction of neuronal differentiation compared with the control. In particular, derivative **26b**, in which an *m*-tolyl group was introduced at the end of the side chain of MK-3, exhibited the highest activity, and its differentiation-inducing effect was about twice that of the control. Based on the ratio of expression levels of Map2 and Gfap, compound **26b** selectively induces differentiation to neurons [28].

Then, compounds **30ab–35ab**, in which heteroatoms were incorporated and substituents such as fluorine and methyl groups were introduced into the phenyl group at the end of the side chain are reported (**Figure 3**). The results of biological evaluation showed that the fluorine-containing compounds enhanced the selectivity of neuronal differentiation. In order to investigate the effect of alkyl

(a)

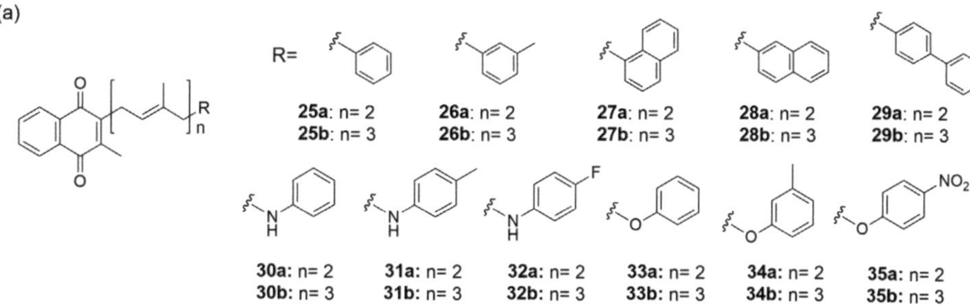

(b)

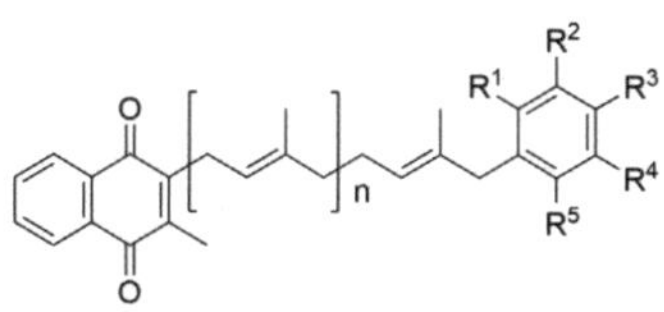

	n	R_1	R_2	R_3	R_4	R_5
36 :	1	CH_3	CH_3	H	H	H
37 :	1	CH_3	H	CH_3	H	H
38 :	1	H	CH_3	CH_3	H	H
39 :	1	H	CH_3	H	CH_3	H
40 :	1	CH_3	H	CH_3	H	CH_3
41 :	1	H	${}^tBu$	H	H	H
42 :	2	CH_3	CH_3	H	H	H
43 :	2	CH_3	H	CH_3	H	H
44 :	2	H	CH_3	CH_3	H	H
45 :	2	H	CH_3	H	CH_3	H
46 :	2	CH_3	H	CH_3	H	CH_3
47 :	2	H	${}^tBu$	H	H	H

Figure 4.
Vitamin K analogs: (a) an aromatic substituent was introduced at the ω-terminal side chain. (b) A tert-butyl group or methyl groups were introduced at the ω-terminal side chain.

groups on the differentiation-inducing activity of compound **26b**, compounds **36–47**, in which several *t*-butyl and methyl groups were introduced into the phenyl group at the end of the side chain, were synthesized. Interestingly, it was clarified that derivatives **36** and **38**, in which two methyl groups were introduced at the 2,3- and 3,4-positions, respectively, of the phenyl group at the end of the MK-2 side chain, inhibited the differentiation of MK-2 into neurons (**Figure 4**) [29, 30].

Thus, the introduction of a hydrophobic functional group at the end of the side chain can enhance the differentiation-inducing activity of vitamin K from neural stem cells to neurons. It is known that natural products such as neuropathiazol, epolactaene, and retinoic acid (retinoid) induce neuronal differentiation. All of these compounds have double bonds or phenyl groups in their side chains, similar to the active vitamin K derivatives synthesized in this study [31–33]. Based on these findings, it might be possible to obtain compounds that have more potent neuronal differentiation activity. At present, the mechanism by which vitamin K derivatives induce neuronal differentiation is unknown. If the proteins upon which vitamin K acts were identified, this would be helpful for rational design of more potent compounds.

As described above, the biological activity of vitamin K is greatly affected by differences in the side chain structure. In addition to vitamin K, many other fat-soluble vitamins, such as vitamins A, D, and E, also have alkyl side chains containing double bonds. This may suggest that there is an optimal side chain structure for each target biological activity, because the specific action of each vitamin differs depending on the alkyl side chain structure. Further investigation of the structure–activity relationships of the side chains and the naphthoquinone part is needed (**Figure 4**).

4. Antitumor activity of vitamin K derivatives

4.1 Menadione-based Cdc25 inhibitors

Antitumor activity is one of the most interesting features of vitamin K and its derivatives. Among synthetic compounds, a series of menadione-based alkylthio naphthoquinone derivatives including 2-hydroxyethylthio-3-methyl-1,4-naphthoquinone (Cpd 5; compound 5, NSC 672121: **48**) are representative examples of vitamin K derivatives with potent antitumor activity. Carr and coworkers designed and synthesized naphthoquinone derivatives bearing an alkyl, alkoxy, or alkylthio group, and evaluated their growth-inhibitory effect toward human hepatoma cell line HepB3. Almost all of the tested compounds, as well as menadione, exhibited significant growth-inhibitory activity toward HepB3 cells, and among the compounds, Cpd 5 exhibited the most potent activity [34]. Further studies revealed that Cpd 5 irreversibly inhibits growth-regulatory phosphatase Cdc25 by arylating the cysteine residue of the catalytic site, causing cell cycle arrest [35–37]. Based on the structure and the mode of action of Cpd 5, various compounds bearing 2-hydroxyethylthio group(s) have been developed as candidate antitumor agents. For example, bis(2-hydroxyethylthio)naphthoquinone derivative NSC 95397 (**49**) shows potent inhibitory activity toward Cdc25 phosphatase and was found to inhibit proliferation of several cancer cell lines [38]. Hydroxylated NSC 95397 derivatives (**50**, **51**) and a fluorinated Cpd 5 derivative (**52**) also exhibited more potent activity than the parent Cpd 5 [39, 40]. A maleimide moiety instead of naphthoquinone could also function as the arylating functionality, and a maleimide derivative PM-20 (**53**) exerted potent growth-inhibitory activity toward HepB3 cells [41].

In addition to the aryl moiety, modification of the sulfide side chain was also investigated. Garbay and coworkers developed carboxylic acid derivatives such as compounds **54**, **56**, and **57**. The carboxy functionality therein was introduced based on the consideration that the carboxylic acid moiety would interact with arginine residues in the catalytic site of Cdc25B, and indeed, these compounds exhibited potent Cdc25B3-inhibitory activity. Though the cytotoxic activity of these carboxylic acid derivatives, especially dicarboxylic acid **57**, was low, benzyl ester derivatives such as **55** and **58**, which could be considered as prodrugs, exhibited enhanced cell growth-inhibitory activity [42, 43]. Suzuki and coworkers investigated the structure–activity relationship of the alkylthio moiety using a series of oxygen-containing derivatives such as **59–61** and found that the methoxy derivative **59** exhibited cytotoxic activity with selectivity toward neuroblastoma cell lines, whereas the parent menadione and Cpd 5 exhibited cytotoxicity toward both neuroblastoma cells and normal cell lines [44].

Because 1,4-naphthoquinone structure as well as quinolinedione structure is considered a promising scaffold for Cdc25 inhibitors, several naphthoquinone-based Cdc25 inhibitors other than Cpd 5 derivatives have been also reported as candidate antitumor agents. Quinolinedione derivatives NSC663284 (**62**) and JUN-1111 (**63**) inhibit Cdc25 function, and the corresponding naphthoquinone derivative **64** also inhibits Cdc25B3 [45, 46]. Recently, Quinn and coworkers synthesized a series of naphthoquinone derivatives and examined their Cdc25-inhibitory activity as well as their binding affinity toward mitogen-activated protein kinase kinase 7 (MKK7). Most derivatives bearing alkylthio group(s) showed both Cdc25-inhibitory activity and MKK7-binding affinity, and NSC95397 (**49**), as well as compounds **65** and **66**, showed marked potency. They also found that compound **67** was a selective inhibitor of Cdc25A/B versus MKK7, whereas compound **68** was a selective inhibitor of MKK7 versus Cdc25A/B [47]. Cdc25 is a promising therapeutic candidate for not only HCC, but also other cancers including triple-negative breast cancer [48]. Development of

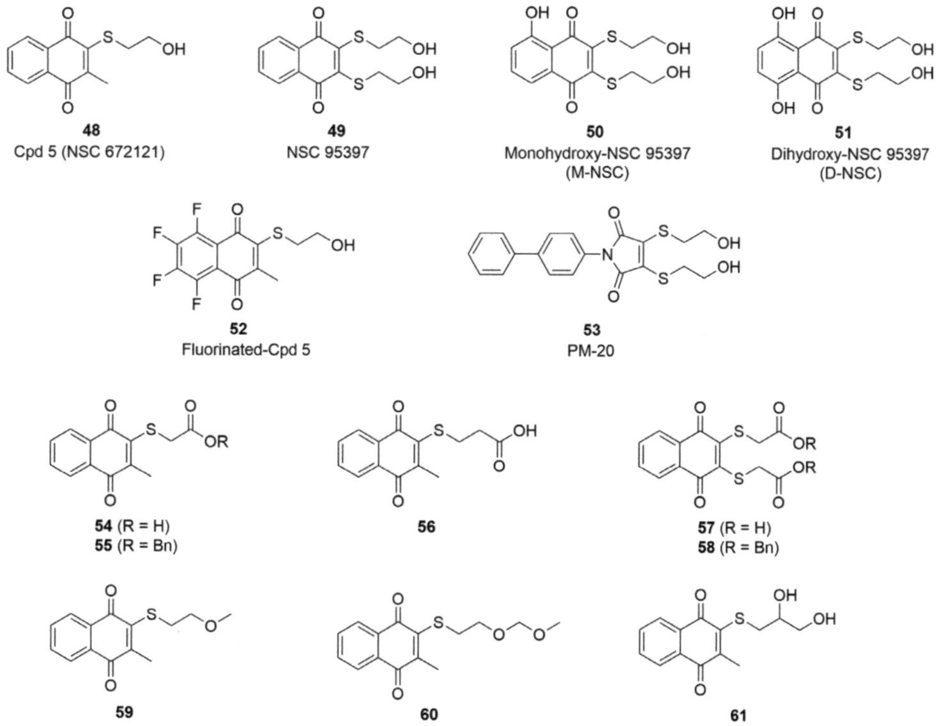

Figure 5.
Structures of Cpd 5 and related derivatives bearing an alkylthio moiety.

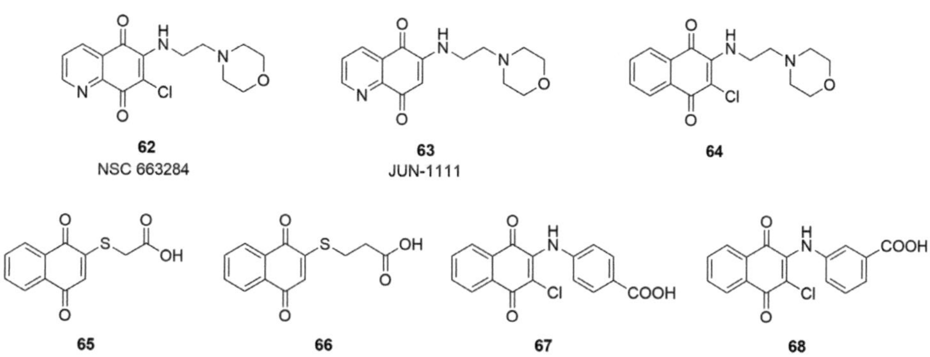

Figure 6.
Examples of naphthoquinone- and quinolinedione-based Cdc25 inhibitors.

vitamin K-based Cdc25 inhibitors could provide novel options for cancer chemotherapy (**Figures** 5 and **6**).

4.2 Anti-hepatocellular carcinoma activity of menaquinone derivatives

The inhibitory effect of menaquinones on tumor progression and the molecular mechanism involved have been intensively investigated [7, 49], and there is continuing interest in the use of menaquinones for the chemoprevention of hepatocellular carcinoma (HCC) due to their safety. Though several clinical studies have suggested a preventive effect of menaquinone against HCC recurrence [50, 51], the efficacy of menaquinones in suppressing HCC was not confirmed in a large-scale clinical study [52]. Therefore, further study of the anti-HCC activity of menaquinones and derivatives is needed. In order to investigate the anti-HCC activity of menaquinones, we focused on carboxylated derivatives, which include isolated and putative metabolites of menaquinones. In the case of MK-4, one of the most interesting vitamin K homologs because of its multifunctional properties, ω-carboxyl homologs of MK-4 (MK-4-ω-COOH: **69**), K acid I (**74**), K acid II (**76**), and their glucuronides, has been identified as metabolites [53–56]. It is considered that MK-4 is metabolized to MK-4-ω-COOH (**69**) by initial ω-oxidation, followed by sequential β-oxidation to afford intermediary carboxylic acids. We focused on the structural similarity between the ω-carboxyl homologs of menaquinones and acyclic retinoid (ACR, peretinoin; **77**), namely a hydrophobic isoprene chain and a carboxyl moiety. ACR, a chemopreventive agent currently under clinical investigation, selectively inhibits HCC cell growth, but also has a limited effect on normal hepatocytes.

Although several synthetic methods for oxidized vitamin K derivatives including K acid I (**74**) and K acid II (**76**) have been reported [57–61], there has been no systematic synthesis of ω-carboxyl menaquinone derivatives. Fujii and coworkers developed a method for systematic preparation of ω-carboxyl menaquinone derivatives using 1,4-dimethoxynaphthalene derivatives instead of reactive 1,4-naphthoquinones as synthetic intermediates [62]. By using the synthesized compounds as standards, McDonald and coworkers newly identified the presence of MK-1-ω-COOH (75) in human urine as a vitamin K metabolite (**Figures** 7 and **8**) [63].

Then, the proliferation-inhibitory activity of ω-carboxyl menaquinone derivatives **70**, **71**, **73**, and **74** toward JHH7 human HCC cells were examined. All the tested carboxylic acid derivatives, including the known vitamin K metabolite K acid I (**74**), exhibited significant proliferation-inhibitory activity toward JHH7 cells, whereas the parent MK-4 had no effect on proliferation. Among the tested compounds, α,β-unsaturated carboxyl derivatives, that is, **71** and **73**, exhibited potent

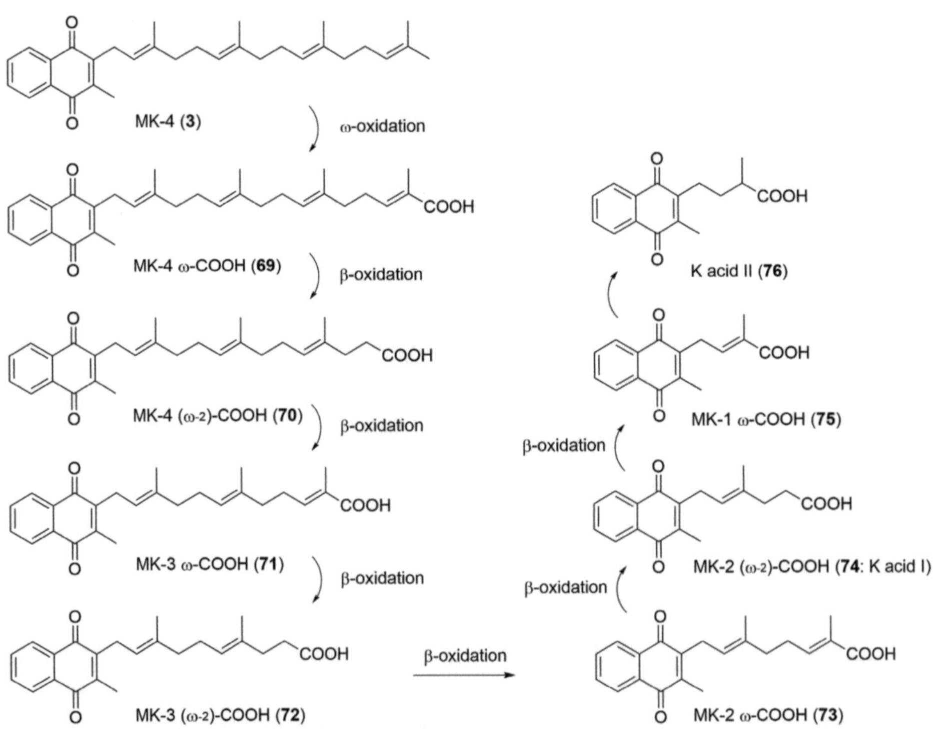

Figure 7.
Putative catabolic pathway of MK-4 based on the identified metabolites.

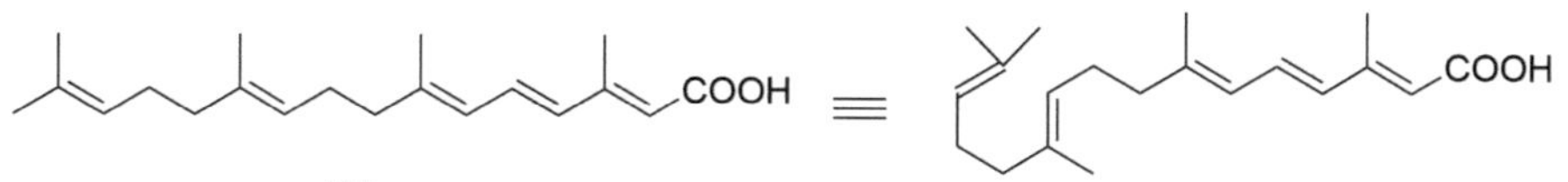

77

Figure 8.
Structure of acyclic retinoid (ACR).

activity. Therefore, the activity profile of the potent compound MK-2-ω-COOH (**73**) was next investigated in detail. Compound **73** inhibited the proliferation of human HCC cell line HepG2, as well as JHH7, but had no significant effect on the proliferation of normal hepatocytes. These results suggested that the growth-inhibitory activity of **73** is cancer-selective. Since it was reported that menaquinone binds to Bak and induces apoptosis of HeLa cells, we next investigated the involvement of Bak in the growth-inhibitory activity of **73**. However, a loss-of-function experiment using siRNA revealed that the observed effect of **73** was not mediated by Bak. As for ACR, it was revealed that a caspase- and transglutaminase-dependent pathway is associated with ACR-induced apoptosis in HCC [64]. Cell growth inhibition caused by compound **73** was reversed by caspase inhibitor ZVAD and transglutaminase inhibitor cystamine, and combined treatment with ZVAD and cystamine almost completely blocked the cell death of JHH7 induced by compound **73**. These results suggested that the proliferation-inhibitory activity of the carboxylated menaquinone derivatives on HCC cells occurs at least partially via caspase- and transglutaminase-dependent pathways [65].

Fujii and coworkers have also developed a different type of candidate anti-HCC agents based on the structure of menaquinones. Specifically, a series of compounds with a phthalazine-1,4-dione core, instead of 1,4-naphthoquinone in the parent menaquinones, and a prenyl substituent corresponding in length to that of MK-1 to MK-4 (**79–82**), were designed. The corresponding ω-carboxylated compounds **83–86** were also synthesized. Phthalazine-1,4-dione is a heterocycle bearing two carbonyl groups, like 1,4-naphthoquinone, enabling us to probe the role of the naphthoquinone moiety in the antitumor effect of the parent menaquinone derivatives. Biological evaluation revealed that the compounds bearing an intact isoprene chain, such as geranyl derivative **80**, exhibited potent anti-proliferative activity toward JHH7 cells, and the growth-inhibitory effects on normal hepatocytes were smaller than those on JHH7 cells. On the other hand, phthalazine-1,4-dione derivatives bearing a ω-carboxylated side chain were mostly inactive toward JHH7, in contrast to the corresponding naphthoquinone derivatives [66]. The SAR of phthalazine-1,4-dione derivatives was different from that of naphthoquinone derivatives. Further investigation of the mechanism of the anti-proliferative effect

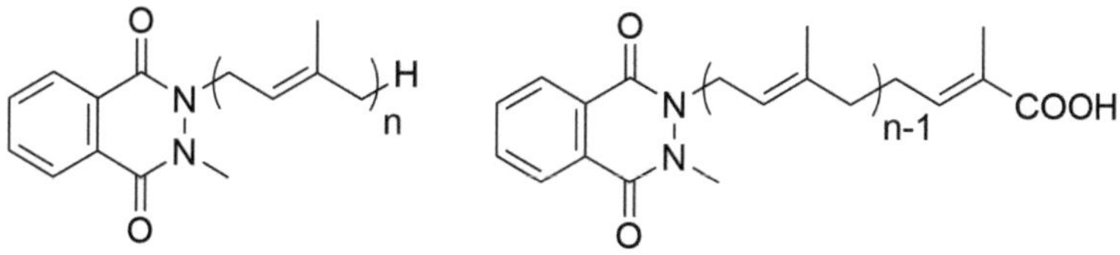

n = 1: **79**
n = 2: **80**
n = 3: **81**
n = 4: **82**

n = 1: **83**
n = 2: **84**
n = 3: **85**
n = 4: **86**

Figure 9.
Structure of menaquinone-based phthalazine-1,4-dione derivatives **79–86**.

of phthalazine-1,4-dione derivatives might provide an improved understanding of the possibilities for chemoprevention of HCC (**Figure 9**).

5. Conclusion

Vitamin K derivatives are attractive lead compounds for drug discovery. In this chapter, three topics in the medicinal chemistry of vitamin K, namely, SXR modulation, neural differentiation, and antitumor effect, were covered. Structure–activity relationship study of menaquinone-based SXR ligands has provided detailed information on the SXR-ligand recognition profile, contributing to the further development of novel SXR modulators. Neuronal differentiation-inducing compounds would be useful as chemical tools to probe signaling pathways that control neuronal specification, and also as candidate therapeutic agents for the treatment of neural diseases. The antitumor activity of vitamin K and its derivatives is also of great interest. Various studies have revealed that Cdc25 is an important target of the antitumor effect of naphthoquinone derivatives, including Cpd 5 and related compounds, and caspase- and transglutaminase-dependent pathways are also potential targets of vitamin K-based anti-HCC agents. Further investigation of the mechanism of the anti-proliferative effect of menaquinone derivatives might lead to agents for the chemoprevention of HCC.

Author details

Shinya Fujii[1], Yoshitomo Suhara[2] and Hiroyuki Kagechika[1*]

1 Institute of Biomaterials and Bioengineering, Tokyo Medical and Dental University, Tokyo, Japan

2 Faculty of Bioscience and Engineering, College of Systems Engineering and Science, Shibaura Institute of Technology, Saitama, Japan

*Address all correspondence to: kage.chem@tmd.ac.jp

References

[1] Litwack G, editor. Vitamin K, Vitamins and Hormones. Vol. 78. New York: Elsevier Academic Press; 2008

[2] Gordeladze JO, editor. Vitamin K2 - Vital for Health and Wellbeing. Rijeka: IntechOpen; 2017

[3] Prasad KN, Edwards-Prasad J, Sakamoto A. Vitamin K3 (menadione) inhibits the growth of mammalian tumor cells in culture. Life Sciences. 1981;**29**:1387-1392

[4] Ngo EO, Sun T-P, Chang J-Y, Wang C-C, Chi K-H, Cheng A-L, et al. Menadione-induced DNA damage in a human tumor cell line. Biochemical Pharmacology. 1991;**42**:1961-1968

[5] Wu FY-H, Chang N-T, Chen WJ, Juan C-C. Vitamin K3-induced cell cycle arrest and apoptotic cell death are accompanied by altered expression of c-fos and c-myc in nasopharyngeal carcinoma cells. Oncogene. 1993;**8**: 2237-2244

[6] Juan C-C, Wu FY-H. Vitamin K3 inhibits growth of human hepatoma HepG2 cells by decreasing activities of both p34-Cdc2 kinase and phosphatase. Biochemical and Biophysical Research Communications. 1993;**190**:907-913

[7] Lamson DW, Plaza SM. The anticancer effects of vitamin K. Alternative Medicine Review. 2003;**8**:303-318

[8] Wu FYH, Liao WC, Chang HM. Comparison of antitumor activity of vitamins K1, K2 and K3 on human tumor cells by two (MTT and SRB) cell viability assays. Life Sciences. 1993;**52**:1797-1804

[9] Miyazawa K, Yaguchi M, Funato K, Gotoh A, Kawanishi Y, Nishizawa Y, et al. Apoptosis/differentiation-inducing effects of vitamin K2 on HL-60 cells: Dichotomous nature of vitamin K2 in leukemia cells. Leukemia. 2001;**15**:1111-1117

[10] Shibayama-Imazu T, Sakairi S, Watanabe A, Aiuchi T, Nakajo S, Nakaya K. Vitamin K2 selectively induced apoptosis in ovarian TYK-nu and pancreatic MIA PaCa-2 cells out of eight solid tumor cell lines through a mechanism different from geranylgeraniol. Journal of Cancer Research and Clinical Oncology. 2003;**129**:1-11

[11] Yoshida T, Miyazawa K, Kasuga I, Yokoyama T, Minemura K, Ustumi K, et al. Apoptosis induction of vitamin K2 in lung carcinoma cell lines: the possibility of vitamin K2 therapy for lung cancer. International Journal of Oncology. 2003;**23**:627-632

[12] Tabb MM, Sun A, Zhou C, Grün F, Errandi J, Romero K, et al. Vitamin K_2 regulation of bone homeostasis is mediated by the steroid and xenobiotic receptor SXR. The Journal of Biological Chemistry. 2003;**278**:43919-43927

[13] Ichikawa T, Horie-Inoue K, Ikeda K, Blumberg B, Inoue S. Steroid and xenobiotic receptor SXR mediates vitamin K_2-activated transcription of extracellular matrix-related genes and collagen accumulation in osteoblastic cells. The Journal of Biological Chemistry. 2006;**281**:16927-16934

[14] Li J, Lin JC, Wang H, Peterson JW, Furie BC, Furie B, et al. Novel role of vitamin k in preventing oxidative injury to developing oligodendrocytes and neurons. The Journal of Neuroscience. 2003;**23**:5816-5826

[15] Ohsaki Y, Shirakawa H, Miura A, Giriwono PE, Sato S, Ohashi A, et al. Vitamin K suppresses the lipopolysaccharide-induced expression of inflammatory cytokines in cultured macrophage-like cells via the inhibition

of the activation of nuclear factor kB through the repression of IKKα/β phosphorylation. The Journal of Nutritional Biochemistry. 2010;**21**: 1120-1126

[16] Spronk HMH, Soute BAM, Schurgers LJ, Thijssen HHW, De Mey JGR, Vermeer C. Tissue-specific utilization of menaquinone-4 results in the prevention of arterial calcification in warfarin-treated rats. Journal of Vascular Research. 2003;**40**:531-537

[17] Otsuka M, Kato N, Shao RX, Hoshida Y, Ijichi H, Koike Y, et al. Vitamin K2 inhibits the growth and invasiveness of hepatocellular carcinoma cells via protein kinase A activation. Hepatology. 2004;**40**:243-251

[18] Habu D, Shiomi S, Tamori A, Takeda T, Tanaka T, Kubo S, et al. Role of vitamin K2 in the development of hepatocellular carcinoma in women with viral cirrhosis of the liver. JAMA. 2004;**292**:358-361

[19] Azuma K, Ouchi Y, Inoue S. Vitamin K: novel molecular mechanisms of action and its roles in osteoporosis. Geriatrics & Gerontology International. 2014;**14**:1-7

[20] Suhara Y, Hanada N, Okitsu T, Sakai M, Watanabe M, Nakagawa K, et al. Structure–activity relationship of novel menaquinone-4 analogues: modification of the side chain affects their biological activities. Journal of Medicinal Chemistry. 2012;**55**:1553-1558

[21] Suhara Y, Watanabe M, Motoyoshi S, Nakagawa K, Wada A, Takeda K, et al. Synthesis of new vitamin K analogues as steroid and xenobiotic receptor (SXR) agonists: insights into the biological role of the side chain part of vitamin K. Journal of Medicinal Chemistry. 2011;**54**:4918-4922

[22] Hirota Y, Tsugawa N, Nakagawa K, Suhara Y, Tanaka K, Uchino Y, et al. Menadione (vitamin K_3) is a catabolic product of oral phylloquinone (vitamin K_1) in the intestine and a circulating precursor of tissue menaquinone-4 (vitamin K_2) in rats. The Journal of Biological Chemistry. 2013;**288**: 33071-33080

[23] Suhara Y, Watanabe M, Nakagawa K, Wada A, Ito Y, Takeda K, et al. Synthesis of novel vitamin K_2 analogues with modification at the ω-terminal position and their biological evaluation as potent steroid and xenobiotic receptor (SXR) agonists. Journal of Medicinal Chemistry. 2011;**54**:4269-4273

[24] Josey BJ, Inks ES, Wen X, Chou CJ. Structure-activity relationship study of vitamin K derivatives yields highly potent neuroprotective agents. Journal of Medicinal Chemistry. 2013;**56**:1007-1022

[25] Moghadam BF, Fereidoni M. Neuroprotective effect of menaquinone-4(MK-4) on transient global cerebral ischemia/reperfusion injury in rat. PLoS One. 2020;**15**:e0229769

[26] Nakayama T, Asami S, Ono S, Miura M, Hayasaka M, Yoshida Y, et al. Effect of cell differentiation for neuroblastoma by vitamin K analogs. Japanese Journal of Clinical Oncology. 2009;**39**:251-259

[27] Eriksson PS, Perlieva E, Bjork-Eriksson T, Alborn AM, Nordborg C, Peterson DA, et al. Neurogenesis in the adult human hippocampus. Nature Medicine. 1998;**4**:1313-1317

[28] Suhara Y, Hirota Y, Hanada N, Nishina S, Eguchi S, Sakane R, et al. Synthetic small molecules derived from natural vitamin K homologues that induce selective neuronal deffirentiation of neuronal progenitor cells. Journal of Medicinal Chemistry. 2015;**58**:7088-7092

[29] Kimura K, Hirota Y, Kuwahara S, Takeuchi A, Tode C, Wada A, et al.

Synthesis of novel synthetic vitamin K analogues prepared by introduction of a heteroatom and a phenyl group that induce highly selective neuronal differentiation of neuronal progenitor cells. Journal of Medicinal Chemistry. 2017;**60**:2591-2596

[30] Sakane R, Kimura K, Hirota Y, Ishizawa M, Takagi Y, Wada A, et al. Synthesis of novel vitamin K derivatives with alkylated phenyl groups introduced at the ω-terminal side chain and evaluation of their neural differentiation activities. Bioorganic & Medicinal Chemistry Letters. 2017;**27**:4881-4884

[31] Warashina M, Min KH, Kuwabara T, Huynh A, Gage FH, Schultz PG, et al. A synthetic small molecule that induces neuronal differentiation of adult hippocampal neural progenitor cells. Angewandte Chemie (International Ed. in English). 2006;**45**:591-593

[32] Kakeya H, Takahashi I, Okada G, Isono K, Osada H. Epolactaene, a novel neuritogenic compound in human neuroblastoma cells, produced by a marine fungus. Journal of Antibiotics (Tokyo). 1995;**48**:733-735

[33] Yu S, Levi L, Siegel R, Noy N. Retinoic acid induces neurogenesis by activating both retinoic acid receptors (RARs) and peroxisome proliferator-activated receptor β/δ (PPARβ/δ). The Journal of Biological Chemistry. 2012;**287**:42195-42205

[34] Nishikawa Y, Carr BI, Wang M, Kar S, Finn F, Dowd P, et al. Growth inhibition of hepatoma cells induced by vitamin K and its analogs. The Journal of Biological Chemistry. 1995;**270**: 28304-28310

[35] Ni R, Nishikawa Y, Carr BI. Cell growth inhibition by a novel vitamin K is associated with induction of protein tyrosine phosphorylation. The Journal of Biological Chemistry. 1998;**273**:9906-9911

[36] Nishikawa Y, Wang Z, Kerns J, Wilcox CS, Carr BI. Inhibition of hepatoma cell growth in vitro by arylating and non-arylating K vitamin analogs. Significance of protein tyrosine phosphatase inhibition. The Journal of Biological Chemistry. 1999;**274**: 34803-34810

[37] Tamura K, Southwick EC, Kerns J, Rosi K, Carr BI, Wilcox C, et al. Cdc25 inhibition and cell cycle arrest by a synthetic thioalkyl vitamin K analogue. Cancer Research. 2000;**60**:1317-1325

[38] Lazo JS, Nemoto K, Pestell KE, Cooley K, Southwick EC, Mitchell DA, et al. Identification of a potent and selective pharmacophore for Cdc25 dual specificity phosphatase inhibitors. Molecular Pharmacology. 2002;**61**: 720-728

[39] Peyregne VP, Kar S, Ham SW, Wang M, Wang Z, Carr BI. Novel hydroxyl naphthoquinones with potent Cdc25 antagonizing and growth inhibitory properties. Molecular Cancer Therapeutics. 2005;**4**:595-602

[40] Kar S, Wang M, Ham SW, Carr BI. Fluorinated Cpd 5, a pure arylating K-vitamin derivative, inhibits human hepatoma cell growth by inhibiting Cdc25 and activating MAPK. Biochemical Pharmacology. 2006;**72**:1217-1227

[41] Kar S, Wang M, Yao W, Michejda CJ, Carr BI. PM-20, a novel inhibitor of Cdc25A, induces extracellular signal-regulated kinase 1/2 phosphorylation and inhibits hepatocellular carcinoma growth in vitro and in vivo. Molecular Cancer Therapeutics. 2006;**5**:1511-1519

[42] Brun M-P, Braud E, Angotti D, Mondésert O, Quaranta M, Montes M, et al. Design, synthesis, and biological evaluation of novel naphthoquinone derivatives with CDC25 phosphatase inhibitory activity. Bioorganic & Medicinal Chemistry. 2005;**13**: 4871-4879

[43] Braud E, Goddard M-L, Kolb S, Brun M-P, Mondésert O, Quaranta M, et al. Novel naphthoquinone and quinolinedione inhibitors of CDC25 phosphatase activity with antiproliferative properties. Bioorganic & Medicinal Chemistry. 2008;**16**: 9040-9049

[44] Kitano T, Yoda H, Tabata K, Miura M, Toriyama M, Motohashi S, et al. Vitamin K3 analogs induce selective tumor cytotoxicity in neuroblastoma. Biological & Pharmaceutical Bulletin. 2012;**35**:617-623

[45] Contour-Galcera MO, Sidhu A, Prévost G, Bigg D, Ducommun B. What's new on CDC25 phosphatase inhibitors. Pharmacology & Therapeutics. 2007;**115**:1-12

[46] Lavecchia A, Di Giovanni C, Novellino E. CDC25 Phosphatase inhibitors: an update. Mini-Reviews Med Chem. 2012;**12**:62-73

[47] Schepetkin IA, Karpenko AS, Khlebnikov A, Shibinska MO, Levandovskiy IA, Kirpotina LN, et al. Synthesis, anticancer activity, and molecular modeling of 1,4-naphthoquinones that inhibit MKK7 and Cdc25. European Journal of Medicinal Chemistry. 2019;**183**:111719

[48] Liu JC, Granieri L, Shrestha M, Wang D-Y, Vorobieva I, Rubie EA, et al. Identification of CDC25 as a common therapeutic target for triple-negative breast cancer. Cell Reports. 2018;**23**: 112-126

[49] Mizuta T, Ozaki I. Hepatocellular carcinoma and vitamin K. Vitamins Hormones. 2008;**78**(Vitamin K): 435-442

[50] Mizuta T, Ozaki I, Egichi Y, Yasutake T, Kawazoe S, Fujimoto K, et al. The effect of menatetrenone, a vitamin K2 analog, on disease recurrence and survival in patients with hepatocellular carcinoma after curative treatment: a pilot study. Cancer. 2006;**106**:867-872

[51] Kakizaki S, Sohara N, Sato K, Suzuki H, Yanagisawa M, Nakajima H, et al. Preventive effects of vitamin K on recurrent disease in patients with hepatocellular carcinoma arising from hepatitis C viral infection. Journal of Gastroenterology and Hepatology. 2007;**22**:518-522

[52] Yoshida H, Shiratori Y, Kudo M, Shiina S, Mizuta T, Kojiro M, et al. Effect of vitamin K2 on the recurrence of hepatocellular carcinoma. Hepatology. 2011;**54**:532-540

[53] Wiss O, Gloor U. Absorption, distribution, storage, and metabolites of vitamins K and related quinones. Vitamins and Hormones. 1966;**24**:575-586

[54] Watanabe M, Toyoda M, Imada I, Morimoto H. Ubiquinone and related compounds. XXVI. The urinary metabolites of phylloquinone and alpha-tocopherol. Chemical & Pharmaceutical Bulletin. 1974;**22** 176-182

[55] McBurney A, Shearer MJ, Barkhan P. Preparative isolation and characterization of the urinary aglycones of vitamin K_1 (phylloquinone) in man. Biochemical Medicine. 1980;**24**:250-267

[56] Tadano K, Yuzuriha T, Sato T, Fujita T, Shimada K, Hashimoto K, et al. Identification of menaquinone-4 metabolites in the rat. Journal of Pharmacobio-Dynamics. 1989;**12**: 640-645

[57] Watanabe M, Kawada M, Nishikawa M, Imada I, Morimoto H. Ubiquinone and related compounds. XXVII. Synthesis of urinary metabolites of phylloquinone and α-tocopherol. Chemical & Pharmaceutical Bulletin. 1974;**22**:566-575

[58] Watanabe M, Okamoto K, Imada I, Morimoto H. Ubiquinone and related compounds. XXXI. Synthesis of urinary metabolites of ubiquinone, phylloquinone, α-tocopherol and related compounds. Chemical & Pharmaceutical Bulletin. 1978;**26**:774-783

[59] Teitelbaum AM, Scian M, Nelson WL, Rettie AE. Efficient syntheses of vitamin K chain-shortened acid metabolites. Synthesis. 2105;**47**:944-948

[60] Okamoto K, Watanabe M, Kawada M, Goto G, Ashida Y, Oda K, et al. Synthesis of quinones having carboxy- and hydroxyl-alkyl side chains, and their effects on rat-liver lysosomal membrane. Chemical & Pharmaceutical Bulletin. 1982;**30**: 2797-2819

[61] Terao S, Shiraishi M, Kato K, Ohkawa S, Ashida Y, Maki Y. Quinones. Part 2. General synthetic routes to quinone derivatives with modified polyprenyl side chains and the inhibitory effects of these quinones on the generation of the slow reacting substance of anaphylaxis (SRS-A). Journal of the Chemical Society, Perkin Transactions. 1982;**1**:2909-2920

[62] Fujii S, Shimizu A, Takeda N, Oguchi K, Katsurai T, Shirakawa H, et al. Systematic synthesis and anti-inflammatory activity of ω-carboxylated menaquinone derivatives – Investigation on identified and putative vitamin K metabolites. Bioorganic & Medicinal Chemistry. 2015;**23**:2344-2352

[63] McDonald MG, Yeung CK, Teitelbaum AM, Johnson AL, Fujii S, Kagechika H, et al. A new LC-MS assay for the quantitative analysis of vitamin K metabolites in human urine. Journal of Lipid Research. 2019;**60**:892-899

[64] Karasawa S, Azuma M, Kasama T, Sakamoto S, Kabe Y, Imai T, et al. Vitamin K2 covalently binds to Bak and induces Bak-mediated apoptosis. Molecular Pharmacology. 2013;**83**: 613-620

[65] Qin X-Y, Fujii S, Shimizu A, Kagechika H, Kojima S. Carboxylic derivatives of vitamin K2 inhibit hepatocellular carcinoma cell growth through caspase/transglutaminase-related signaling pathways. J Nutrit Sci Vitaminol. 2015;**61**:285-290

[66] Fujii S, Miura T, Oikawa T, Qin X-Y, Kojima S, Kagechika H. Design, synthesis and antitumor activity of phthalazine-1,4-dione-based menaquinone analogs. Bioorganic & Medicinal Chemistry Letters. 2021;**43**:128065

Chapter 2

A Perspective of Diverse Synthetic Approaches and Biological Applications of Vitamin K

Satyanarayana Battula

Abstract

Vitamin-K is a demanding multi-functional health product in the market and belongs to a class of isoprenoid molecules that comprises methylnaphthoquinone (MK) unit attached to an isoprene side chain. They are fat soluble and differ in the extent of side chain & obtained in the nature as vitamin K_1 (phylloquinone), menaquinone/vitamin K_2, and other lipoquinones. Owing to their owned polyprenyl side chain, they are hydrophobic/lipophilic in nature. Generally, the synthesis of vitamin K and its variants suffers with isomerization (for example 11 isomers were identified for cis/trans MK-7). Naturally, in bio-systems vitamin K produces through shikimic acid pathway and terpene biosynthetic pathway for the synthesis of menaquinone part & prenyl side chain parts respectively. Menadione or its auxiliaries are commonly being used as substrates to the synthesis of vitamin K variants through the involvement of condensation reactions, Friedel-Craft alkylation's, Claisen rearrangement, Diels-Alder reactions and others. Importantly, organometallic reagents, such as Grignard, Gilman, organotelluride and other reagents could be the promising and consistent choice of substrate to the synthesis of various vitamin K's. Vitamin K is well known for blood coagulation. As an antihaemorrhagic vitamin, it's also being the current interest for the treatment of bone and vascular diseases. In addition, vitamin k is indispensable for the activation of vitamin K dependent (VKD) proteins and that are present almost in all tissues and responsible for hemostasis, bone mineralization, arterial calcification, apoptosis, phagocytosis, growth control, chemotaxis, and signal transduction. This chapter summarizes various synthetic approaches of vitamin K & derivatives and their biological functions.

Keywords: Vitamin K, Menaquinones, Biosynthesis, Synthetic auxiliaries, Organometallic reagents, MenJ, Vascular diseases, diabetes

1. Introduction

Vitamin K is a family of natural products, comprises vitamin K_1 (phylloquinone), vitamin K_2 (methylnaphthoquinones/menaquinones-MK) and vitamin K_3 (menadione). These are structurally methylated napthoquinones possess isoprene side chain (vitamin K_1 & vitamin K_2) and they vary in the extent of isoprene side chain, in terms of number of isoprene units and its level of saturation [1–5]. However, these structural changes are apparently simple, but needs

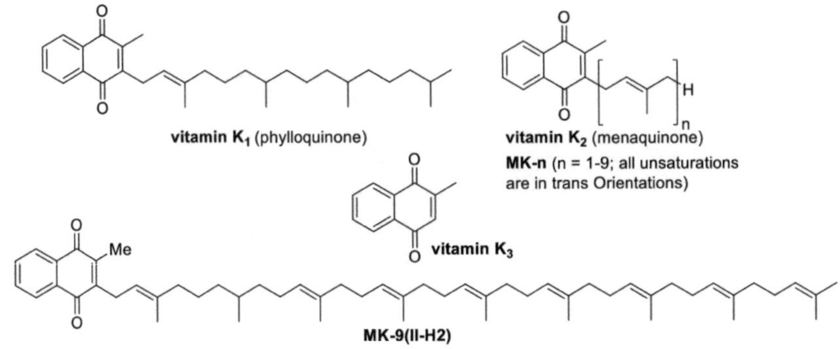

Figure 1.
Structures of vitamin K.

exist in its specific stereochemistry to exhibit their biological functions. These are the fat-soluble compounds and pertains to a class of lipoquinones [6]. Vitamin K_2 are varied on the basis of number of isoprene units present in the side chain and they denoted as MK-n (n – number of isoprene units; **Figure 1**). The vitamin K_2 if contains saturated isoprene units referred as MK-n(m-H_2), wherein m is a roman numeral represents isoprene unit number in the side chain which underwent reduction. Among the two natural forms of vitamin K, vitamin K_1 presents in green leafy vegetables, for example, kale, collard greens, turnip greens, iceberg lettuce, broccoli, spinach, and brussels sprouts. The Vitamin K_2 presents naturally in eggs, meet, fermented foods (natto, cheese, yogurt and sauerkraut) also present in bacteria, for example, MK-9 is found in mycobacterium with nine isoprene units and also its reduced derivative at second isoprene unit MK-9(II-H_2) could be active as electron transport agent in [3, 7]. In humans, these menaquinones display several biological properties, including facilitating blood coagulation [8, 9]. In bacteria, these molecules assisting the synthesis of ATP through transport of electrons between the membrane-bound protein complexes and thus acting as electron acceptors and donors in the respiratory electron transport [10, 11]. Vitamin K and its analogues could not synthesize by mankind/animals. So, its required to supply an adequate amount through the dietary sources.

2. Vitamin K biosynthesis

Naturally, bacteria producing different variants of vitamin K. Among the aerobic bacteria, most of its Gram-negative bacteria contain ubiquinone as the sole quinone, whereas the menaquinone is the only quinone presents in aerobic Gram-positive bacteria. But, in the case of anaerobic bacteria, irrespective of whether it is Gram-positive or Gram-negative, it produces benzoquinones (ubiquinone), naphthoquinones (menaquinones; MK-n), demethylmenaquinones (DMK-n) [12]. The Gram-negative bacteria, such as, *Escherichia coli* and *Salmonella enterica* serovar Typhimurium possesses these isoprenoid quinone molecules, viz., benzoquinones and naphthoquinone. In *E. coli*, it contains MK-8 and DMK-8 as the major naphthoquinones & in addition, it also comprises MK-6, MK-7, MK-9 and DMK-9 in minor quantities (**Figure 2**) [13].

These menaquinones have prenyl side chains with all-*trans* configurations. MK-10 and MK-11 are the isoforms of vitamin K_2 and are produced from bacteroides. MK-7 produced from veillonella sp. and MK-6 is from *Eubacterium lentum*. The menaquinone nucleus of vitamin k was being synthesized in the nature by shikimic acid pathway and that can be described as a "metabolic tree with many branches

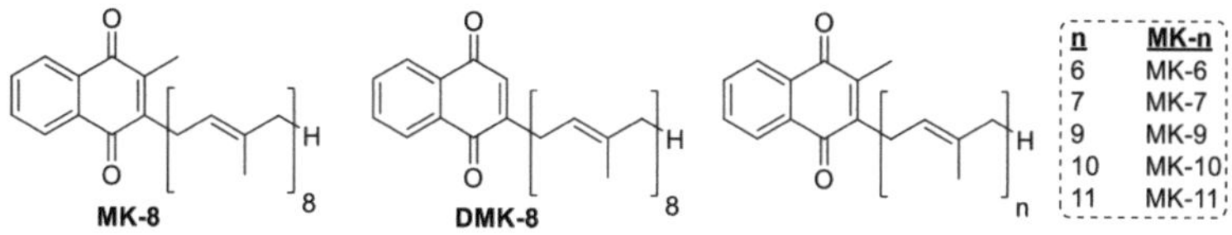

Figure 2.
Structures of vitamin K_2 (MK-n) & DMK-n.

and was derived from isochorismate via chorismate. The prenyl side chains in the molecules are synthesized before its incorporation into the final compound through the terpene biosynthetic pathway and are derived from prenyl diphosphate, and the methyl groups in the vitamin K are derived from S-adenosylmethionine. The menaquinone biosynthesis involves the introduction of prenyl side chain, and that is accompanied by the loss of the carboxyl group and then occurred the C-methylation on quinone moiety [12].

Cox and Gibson discovered in 1964, shikimate was present in menaquinone of *E. coli*, and thus it could be the first report as an evidence to shikimate pathway to the synthesis of menaquinones. The chemical degradation reaction of labeled isolated MK-8 displayed that all the radioactivity is associated with degraded product phthalic anhydride. It indicated that the benzene ring of naphthoquinone part in the vitamin K_2 was procured from shikimate in the bacteria [14]. It was further established that, the menaquinone biosynthesis was started with shikimate as a starting precursor and it was confirmed from a complete degradation study of the menaquinone which was constructed by the labeled shikimate molecule. This study was indicated that all the seven carbon atoms of shikimate were incorporated in the naphthoquinone part of menaquinone [15], and the remaining 3 carbon atoms were revealed to be derived from the 2-ketoglutarate (**Figure 3**) [16, 17].

Shikimate pathway for the synthesis of menaquinones is presented in **Figure 4**. Shikimate was proposed to converted initially into chorismite before its incorporation into menaquinone [15]. In *E. coli*, isochorismate formation occurred from chorismite and it was mediated by isochorismate synthase (encoded as *MenF*). The source of the hydroxyl group in the isochorismate could be either molecular oxygen (occur in aerobic organism), or intramolecular -OH group transfer or from the solvent water [18]. Later, it was postulated that, the 2-ketoglutarate gives succinic semialdehyde anion of TPP through a TPP-dependent decarboxylation and that catalyzes by 2-ketoglutarate decarboxylase (KDC) [16, 19]. Subsequently, this anion reacts with isochorismate and produces 2-succinyl-6-hydroxy-2,4-cyclohexadiene-1-carboxylate (SHCHC). MenD protein was found that it incorporated both SHCHC synthase & KDC activities and was encoded with a single gene [20]. Further studies

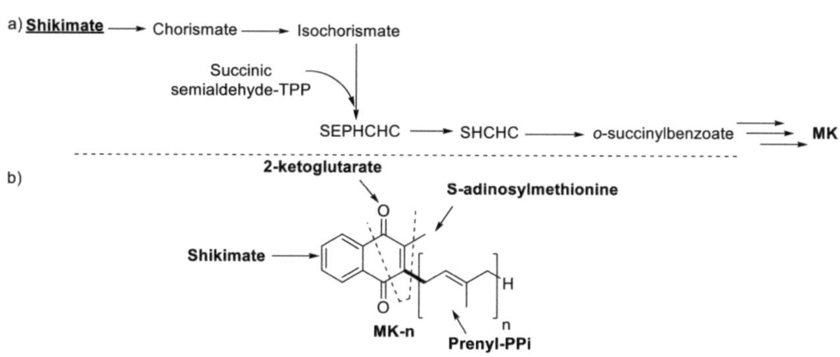

Figure 3.
a) Sequential formations in the biosynthesis of menaquinones; b) the preliminary precursors in MK-n biosynthesis.

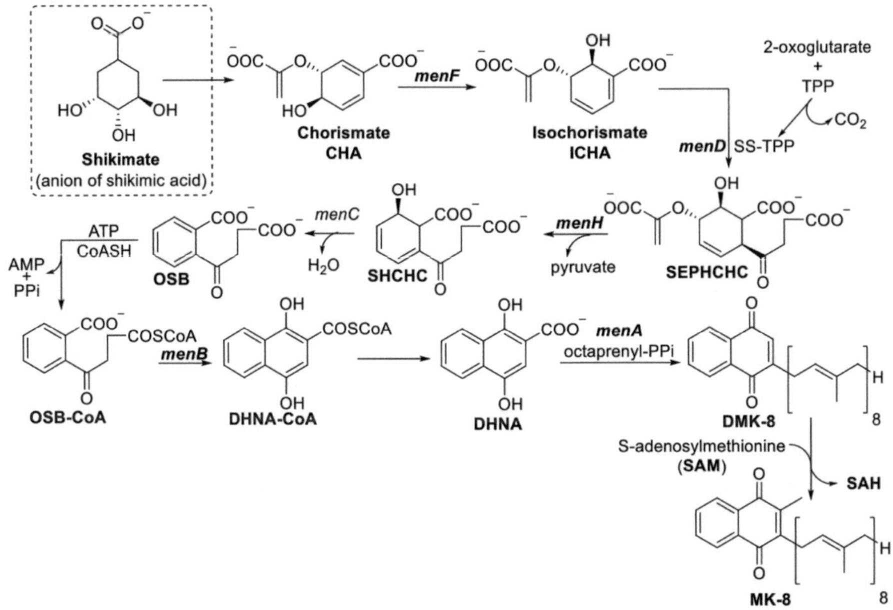

Figure 4.
Biosynthetic pathway of Menaquinone molecules.

shown that, SHCHC formation is proceeded through the formation of 2-succinyl-5-enolpyruvyl-6-hydroxy-3-cyclohexene-1-carboxylate (SEPHCHC) and was catalyzed by SEPHCHC synthase [21]. As this SEPHCHC compound unstable, under mild basic conditions, it converts to SHCHC spontaneously [22]. This *in vivo* conversion was carried out by SHCHC synthase encoded by *menH* gene (Ser-His-Asp; a catalytic triad) [23]. Later, the removal of water molecule from SHCHC to produces benzenoid compound, *o*-succinylbenzoate (OSB) and this transformation was carried out by OSB synthase & encoded as *menC* [24]. Bryant and Bentley described the transformation of OSB in to the 1,4-dihydroxy-2-naphthoate (DHNA) [25]. It was derived by ATP and CoA, so that OSB-CoA could be the proposed intermediate driven by the *menE* gene. Two enzymatic activities are worked on this conversion, viz., OSB-CoA synthetase and DHNA synthase. Due to the un-stability of OSB-CoA, as an intermediate it was converted to DHNA through the involvement of DHNA-CoA [26]. Further Bentley was shown that the conversion of DHNA to DMK was happened in the extracts of *E. coli* [27]. The transformation of DHNA to DMK occurs through the exchange of carboxyl with isoprenpoid side chain. Both these processes, prenylation and decarboxylation perhaps occur at a single reactive center [28]. The *MenA* enzyme has carried out this process with octaprenyl PPi [29, 30]. Finally, the conversion of DMK to MK was happened by a methyltransferase, and that uses *S*-adenosylmethionine as the methyl donor [25, 31]. Moreover, Meganathan and group was extensively studied and reviewed detailed mechanistic studies for the shikimate pathway mechanism for the biosynthesis of vitamin K_2 [12, 32].

3. Various synthetic approaches to vitamin K

In 1939, Fieser [33], Binkley [34], and Almquist [35] were reported initially the synthesis of vitamin K_1 independently. The condensation reaction of either menadione/2-methyl-1,4-naphthohydroquinone with natural phytol in the presence of oxalic acid or zinc dust in acetic acid produce vitamin K_1 (**Figure 5**).

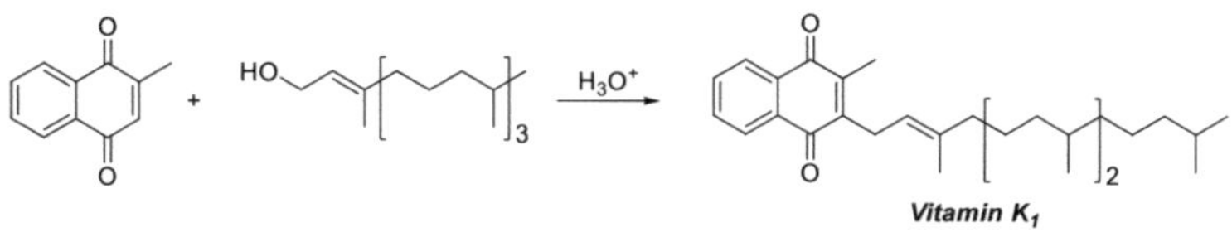

Figure 5.
The first synthesis of vitamin K_1.

The initial synthetic approaches were generally proceeding through the usage of Friedel-Craft alkylation's for introducing the side chains through its coupling to menadions, which led to the generation of mixture of isomers at the Δ_2 position and produced *E*-isomer as the major (90%). Later it was developed by Lindlar to accomplish the complete retention of configuration of the side chain by taking the use of menadiol ester (menadiol 1-benzoate) in the catalytic amount of $BF_3.Et_2O$ instead of menadione [36]. But the reaction suffers with the allylic alcohol instability to the acidic reaction conditions. Later days, the synthesis of vitamin K was achieved by coupling reactions of protected menaquinones with side chains, the protection is depending on the nature of the substrates and deprotection protocols. Transition metals (Ni [37], Ag [38]) show huge applicability in these coupling reactions to the synthesis of vitamin K. Furthermore, organometallics did this synthesis through the reactions of their metallonaphthalenes with corresponding side chain halides. Among, Grignard reagents were given excellent reactions with 99% stereo retention of the *E*-configuration. By using this magnesium organometallics Evans and Hoffman did a regiospecific isoprenylation of naphthoquinones to the vitamin K [39]. In this reaction, trimethylsilyl cyanide (TMSCN) added 2-methoxy-3-methyl-1,4-naphthoquinone gave the protected quinone wherein cyanide was acted as a catalyst. This protected quinone underwent reaction with prenyl magnesium bromide followed by a Cope rearrangement to obtained the protected MK-1. It's deprotection was later performed with AgF, consequently it afforded the MK-1 (**Figure 6**).

In addition, Organocuprates were also being used to the synthesis of vitamins K's. Menaquinone and phylloquinone are synthesized by Chenard group, from the reaction of Gilman based bisketals of quinone substrate with allyl halides and it afforded high yields and good stereoselectivity at the Δ_2 position of the vitamin K (**Figure 7**). The quinone bromide underwent metalation to produce the corresponding cuprate, and its reactivity was varied with different electrophilic substrates (RX). Among the tested electrophilic substrates, if the electrophilic group is small enough (allyl bromide) then the two alkyl groups of cuprate reagent were being used being transferred in the reaction. If the reaction with bulkier electrophilic halide (for example benzyl chloride/bromide, cyclohexanecarbonyl acid chloride), cuprate can transform only one alkyl group [40]. Syper group synthesized protected forms of MK-1 and MK-2 in appreciable yields through the coupling reaction of prenyl bromide and geranyl bromide with 2-bromo-3-methyl-1,4-dimethoxynaphthalene [41]. Generally, these organometallics (Grignard reagents, organocuprates and organolithiums) mediated synthesis of vitamin K required the usage of protected quinones to avoid the side reactions. Unprotected quinones could also give the vitamin k synthesis through their direct coupling with organostannates [42] and organisilanes [43].

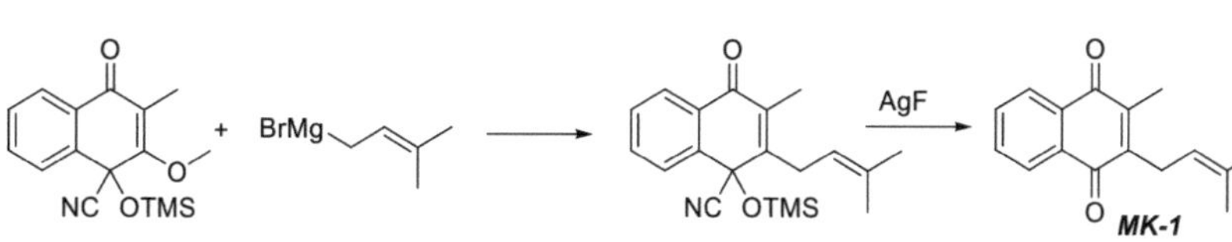

Figure 6.
Grignard reagent mediated synthesis of MK-1.

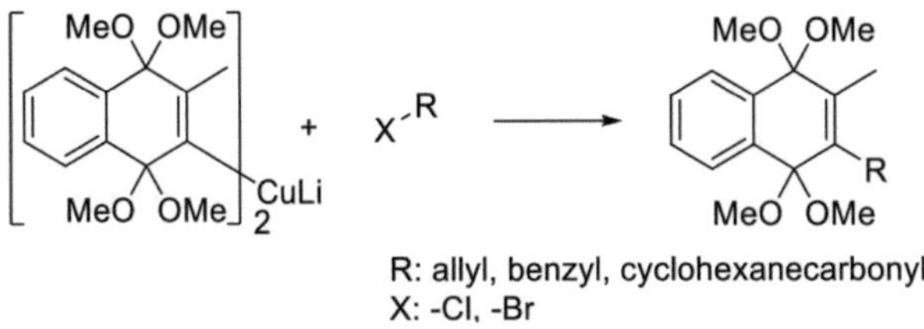

Figure 7.
Menaquinone analogs synthesized by organocuprates.

Tso and group developed an efficient and conceptually distinguished one-pot protocol to the synthesis of vitamin k, wherein the 3-substituted isobenzofuranone was treated with a base, the generated quinone methide which underwent an anionic [4 + 2] cycloaddition reaction with the alkenyl phenyl-sulfone (dienophile) and that was being synthesized from the corresponding allyl phenylsulfone and various prenyl bromide (RBr). Finally, the vitamin K was produced by the elimination of sulfone from the intermediate. This method was very compatible to the synthesis of phylloquinone and different menaquinone variants MK-1, MK-2 and MK-9 about 60–65% yields (**Figure 8**) [44].

In 2015, Mal et al., extended this protocol to the synthesis of menadione derivatives (MK-n molecules) by a base mediated reaction of 3-substituted phthalide with methyl methacrylate [45]. The yield of the reaction was verified in this reaction with various leaving groups, such as - methoxy, phenyl sulfonyl, isonitrile, thiophenyl, and nitrile. The nitrile leaving group in the substrate was found to be the best suitable one for the reaction. The 3-nitrilephthalide is to be a menadione auxiliary to the reaction of MK-n variants (**Figure 9**). The reaction was proceeds through a base mediated anionic driven annulation of 3-nucleofugal phthalides with α-alkyl/aryl acrylates followed by demethoxycarbonylation. If the polyprenyl acrylates are to be the substrates, then various analogs of menadione were being produced (MK-n).

Side chain functionalization methods were also being developed to the synthesis of vitamin K analogs. These derivatized vitamin K are received a great deal of interest to reveal the structure activity relationship studies (SAR). These analogs were also proved to be as inhibitors of vitamin K dependent carboxylase and vitamin K epoxide reductase [46]. The side chain stereochemistry is very essential to exhibit the biological activity, as the *E*-isomers of these vitamins are producing activity and the isomers with *Z*-configuration could lead to loss of their activity. The longer chain in the menaquinones causes to account lower biological activities due to their hydrophobicity [47].

Snyder and group during their sustained efforts to retain the stereochemistry of *α*-isoprene double bond, they introduced the alkylation of enolates to the synthesis of vitamin K [38]. The synthesis involves the interaction of nucleophilic aromatic component to the receptive prenyl component through its enolized nucleophile.

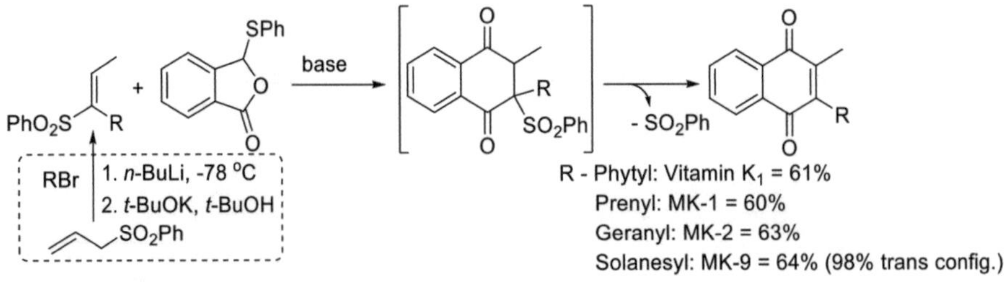

Figure 8.
Methide annulation to a one-pot synthesis of vitamin K.

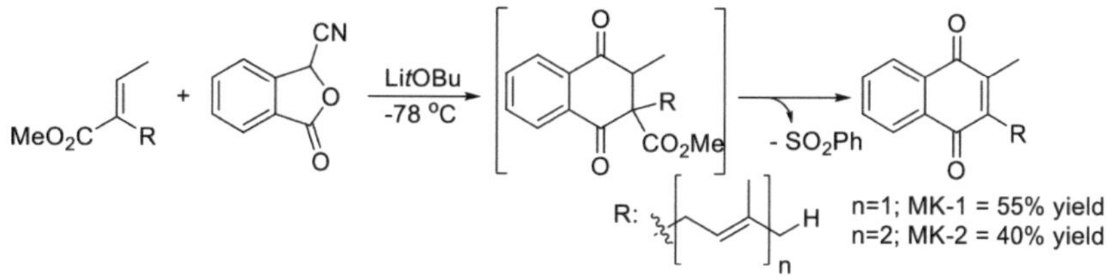

Figure 9.
Anionic annulation of 3-nucleofugal phthalides to the synthesis of vitamin K.

Initially, the menadiol forms its potassium salt in presence of either potassium hydride or potassium methoxide and then underwent prenylation through the enolate alkylation, later in presence of silver oxide get oxidized into prenyl fragment substituted menadione (MK-2 and MK-9) and was shown in **Figure 10**.

However, these enolyte alkylations were successful but not very practical. To accomplish the practical methods, Snyder group developed transmetallation method for the synthesis of vitamin K [38]. Lithium, magnesium and copper had used to convert 2-bromomenadiol to its corresponding metallo derivative. As an electrophilic partner authors chosen initially aldehydes, but aldehydes are unsuccessful as the resultant alcohol functionality removal afforded either vinyl alkenes or α-isoprene double bond isomerization. Later, the prenyl halides are found to be the appropriate substrates of electrophiles for these transformations. Primarily, reactions were performed to evaluate the stereoretention of the α-isoprene double bond in each of the 2-metallo derivatives. As per their expectation, all these metallo derivatives could not affect the configuration of the isoprene component of the vitamins. It was found that, organo magnesium reagents are more efficient than organocuprates and that in turn efficient than organo lithium reagents. Among the prenyl halides, prenyl bromides are generated excellent yields of menaquinone analogs (**Figure 11**). The coupling of 2-magnesio or 2-cupro-3-methyl-1,4-dimethoxynaphthalene with geranyl bromide produced in >90% yield. In this method, MK-9 was obtained in 73% yield by treating solanesyl bromide with menadiol.

In the various synthetic strategies to the vitamin K in the literature, Friedel-Crafts alkylation protocol of menadiols is the most popular across the literature. Hirschmann group developed a method wherein the applicability of various Lewis and Bronsted-Lowry acids were being used for the synthesis of vitamin K_1 [48]. Monoacetate of menadiol substrate was treated with phytol in presence of various acids. Among, the acids, $BF_3.Et_2O$ afforded appreciable yields (67%) rather others being tested $KHSO_4$, oxalic acid and Duolite C-60. After Friedel-Crafts alkylation, finally, the acetate group was removed by the treatment with Ag_2O (**Figure 12**).

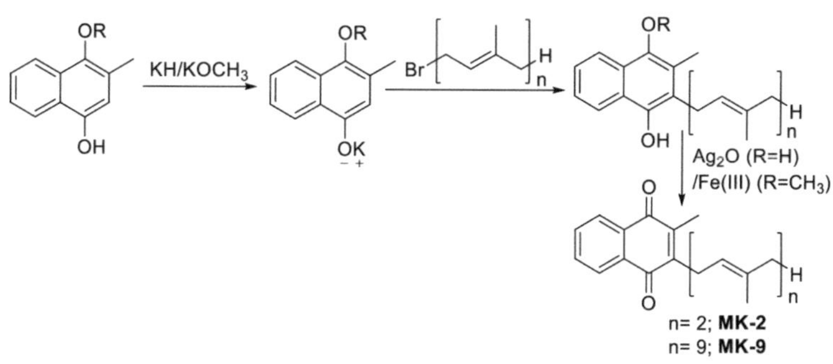

Figure 10.
Enolate alkylation of menadiols to MK-n synthesis.

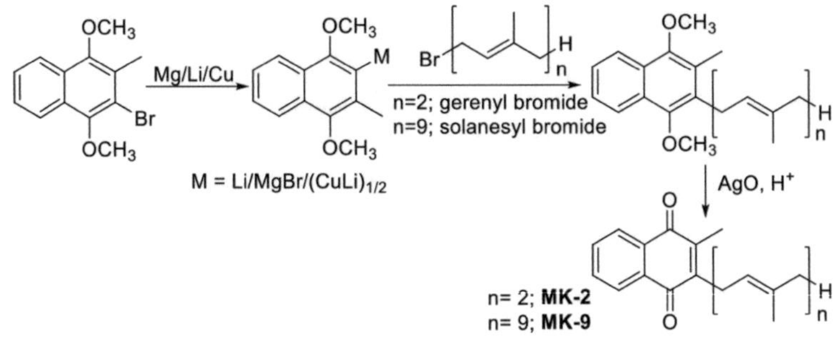

Figure 11.
Efficient transmetallation reaction to menaquinones.

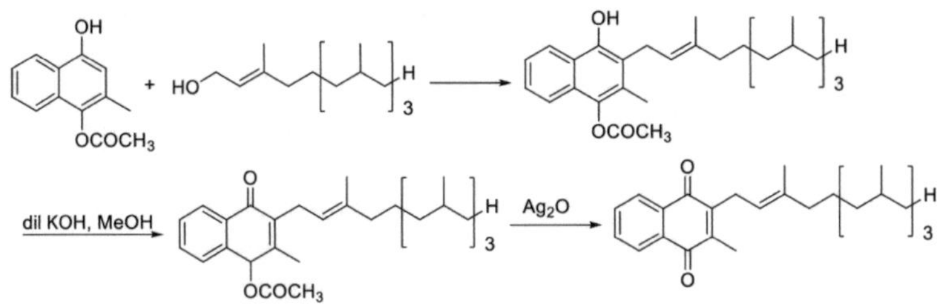

Figure 12.
Application of Friedel-crafts alkylation to the synthesis of vitamin K_1.

Later, in 1990 Schmid and group enriched the yield of the reaction by treating the reaction of menadiol monoacetate with phytyl chloride in presence of a base potassium carbonate, wherein O-phytylated derivative had obtained. This was further undergoing Claisen rearrangement under acidic treatment with catalytic amount of $BF_3.Et_2O$ to C-phytylated product in 76–80% yield and the ratio of the configurations of E and Z isomers was found to be 97:3. The deacetylation was carried out in basic medium (**Figure 13**). Vitamin K_1 obtained in this procedure is to be 96.5% and with E-configuration, Z -configuration compound was with very low yield (3%) [49].

Apart from the electrophilic side chain attachment to the menadiones, radicals also found to do the same work to make the synthesis of vitamin K. Jacobson and coworkers in 1972, developed a method to the alkylation of quinones by the use of radicals and that were produced by metal/persulphate catalyzed decorboxylation of the corresponding carboxylic acid. 4-Methyl-3-pentenoic acid produced the 3,3-dimethylallyl radical while in presence of $AgNO_3$, $(NH_4)_2S_2O_8$. Initially Ag^+ is reacted with $S_2O_8^{-2}$ to form Ag^{+2} and which abstract an electron from carboxylic

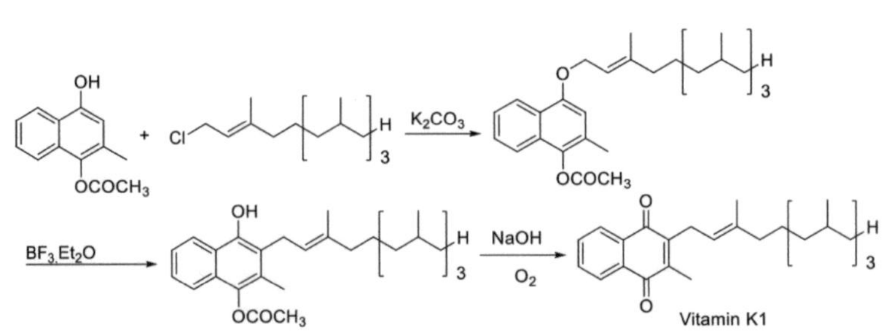

Figure 13.
BF3.Et$_2$O catalyzed Claisen rearrangement to vitamin K_1.

acid to produce allyl radical by the ejection of CO_2. This allyl radical resonates and unexpectedly less stable isoform γ,γ-dimethylallylquinone was given the product instead to α,α-dimethylallylquinone (**Figure 14**). By this procedure MK-1 obtained in 70% yield as a γ,γ-dimethylallylquinone as to hold a stable alkene [50].

Later, Yamago et al., found the applicability of organotellurium compounds to the radical coupling reactions with a variety of quinone substrates [51]. This method could offer the general protocol to the synthesis of polyprenyl menadiones with complete retention of stereochemistry. Organotellurides were prepared by SmI_2 mediated coupling reaction of its corresponding bromides with ditollyl ditelluride $(Tol)_2Te_2$. This telluride later produced prenyl radical under photo/thermal energy source and that interact with menadione to produce respective MK-n related to the length of the prenyl side chain. While geranyl tolyl telluride (73:23 mixture of the trans and cis isomers) react with 2-methyl-1,4-naphthoquinone to produce MK-2 in moderate yield. The stereochemistry of the product (7:3) informed that the reaction gave the retention of stereo chemistry of organic telluride in to the product (**Figure 15**).

Diels-Alder reactions were useful to construct napthoquinine structural unit, as Rüttimann and group being used this concept to develop protocol for the synthesis of vitamin K_1. The reaction of dihydroisobenzofurane was performed with activated alkyne dienophile (96:4 E/Z) at 80°C overnight to form the trimethyl sillyl ether of Diels-Alder adduct and its reaction further carried out with methanol and then methyl group at C_2 position is achieved through the reduction of ester with bis(2-methoxyethoxy)aluminum hydride followed by air oxidation produces vitamin K in moderate yield (**Figure 16a**, 50%) [52]. During the reaction the isoprene double bond configuration was not altered. Later, he developed the reaction by taking an auxiliary to support the reaction. Rüttimann along with Büchi had taken cyclopentadiene as a substrate auxiliary and its reaction was carried out with menadione to generate Diels-Alder adduct (**Figure 16b**). Prenylation/ alkylation at C3 position of Diels-Alder adduct was performed under strong base (KO*t*Bu, $NaNH_2$, KNH_2).

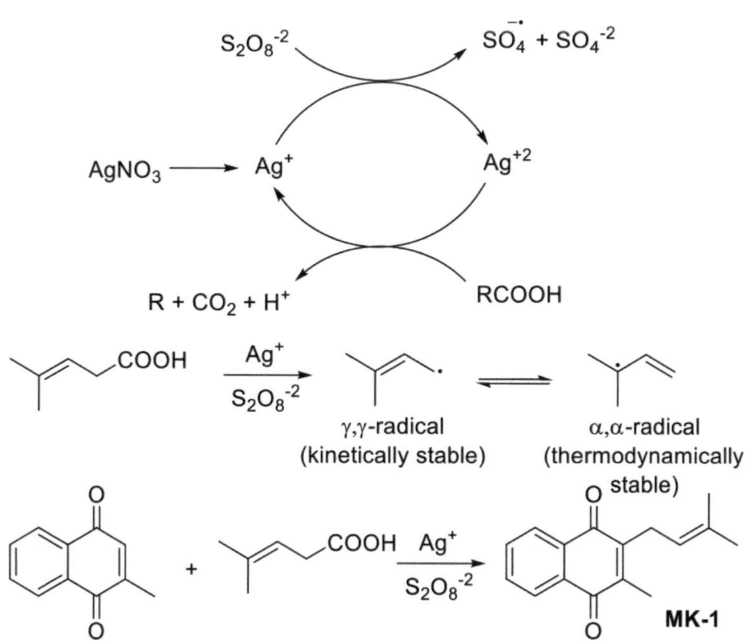

Figure 14.
$Ag^+/S_2O_8^{-2}$ mediated radical directed synthesis to MK-1.

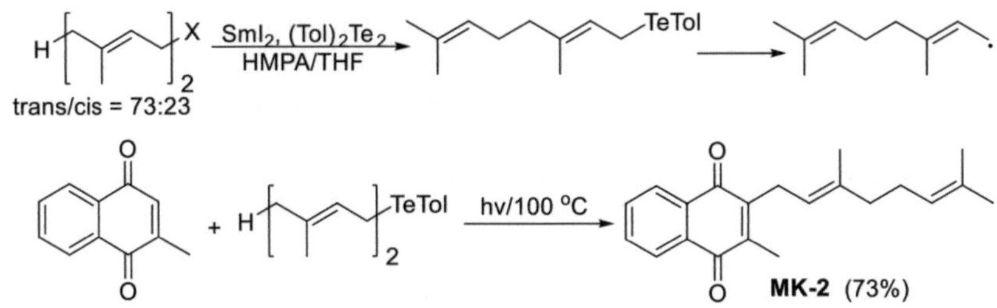

Figure 15.
Organo telluride mediated radical coupling synthesis to vitamin K.

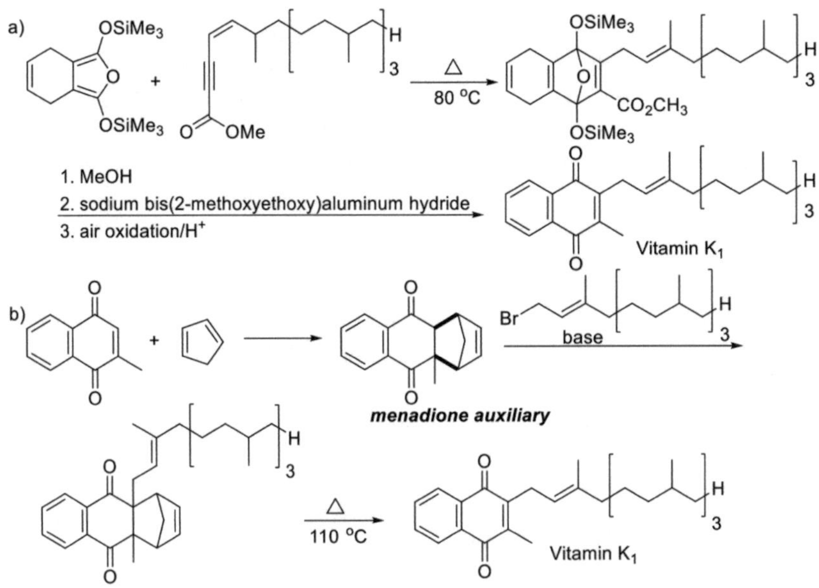

Figure 16.
a) Diels-Alder reaction mediated synthesis to vitamin K_1; b) Cyclopentadiene auxiliary driven synthesis to vitamin K_1.

Inspired by menadione auxiliary applicability in the synthesis of vitamin K_1, Battula, S., and group, thought to introduce the polyprenyl side chain on to the menadione to synthesis MK-n variants. This menadione surrogate was utilized to the synthesis of MK-9 in one pot protocol. The reaction of 1-Chloro-*N,N*,2-trimethyl-1-propenylamine (Ghosez reagent) with solanesol and Diels-Alder adduct of menadione was produced MK-9, wherein solanesol chlorination was happened initially to produce solanesyl chloride and that was treated with menadione auxiliary in presence of a strong base KO*t*Bu. Finally, the reaction was treated with acetic acid followed by tributylmethylammonium bromide to eliminate the by-products. In this procedure, MK-9 was obtained in 77% yield (**Figure 17**). The reaction with bromine based Ghosez reagent was generated MK-9 with 65% yield, as its higher leaving group ability than chloride facilitates to the formation of side reactions like $S_n^{2'}$ reaction and cyclic reactions [53].

In the view of the requirement of MK-7 owing to its high lipophilicity and good bioavailability in small intestines and about 3 days half-life than compare to other menaquinones and ubiquinone, Aneta et al. in 2016 developed a practical synthetic strategy for vitamin K_2 (MK-7) [54]. The synthesis of MK-7 was achieved in all the *trans* forms in the side chain through 1 + 6 convergent synthetic protocol and that involves condensation of menadione monoprenyl fragment with all-*trans* hexaprenyl

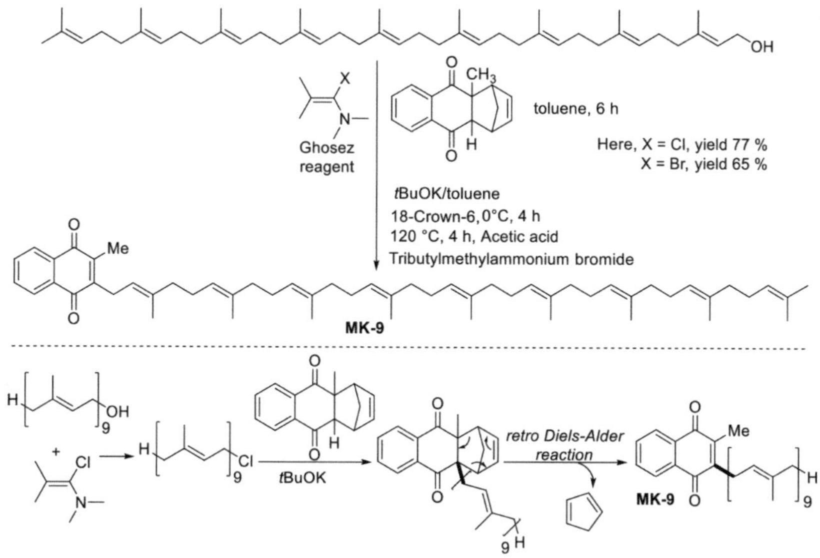

Figure 17.
One-pot synthesis to MK-9.

bromide fragment. All-*trans* hexaprenyl bromide fragment was being synthesized from two tryprenyl molecule, i.e., *trans,trans*-farnesol. This method also implemented to the industrial level synthesis of MK-7 to compensate the need of dietary supplement as it affords high purity of the final compound MK-7 (**Figure 18**).

Among the two coupling components, hexaprenyl fragment was obtained from commercially available *trans,trans* farnesol & *trans,trans* farnesyl bromide by the following synthetic steps. The other substrate monoprenyl 1,4-dimethoxy naphthalene was produced from menadione and phenylsulfonyl, hydroxy isoprene compounds. Both these substrates were coupled in the presence of strong base

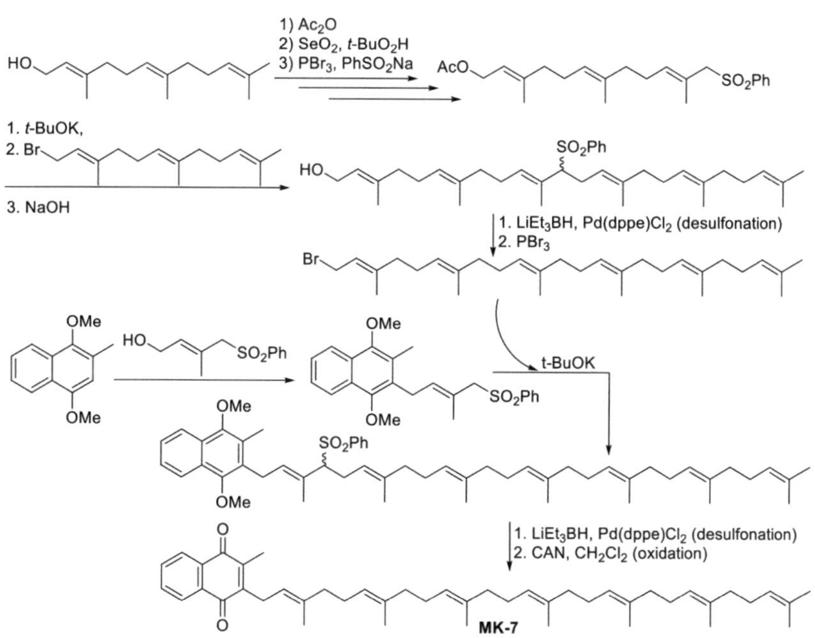

Figure 18.
Convergent synthesis of MK-7.

t-BuOK, followed by the desulfonation at the side chain was performed by $LiEt_3BH$ in presence of $Pd(dppe)Cl_2$. Later, the 1,4-dimethoxy naphthalene was oxidized with CAN. This "1 + 6" convergent synthetic strategy is based on the condensation of monoprenyl derivative of menadione with hexaprenyl molecule was produced MK-7 in all trans stereo isoforms of side chain.

As this method was convenient in the availability of starting substrates and perfect stereochemistry of the product, Battula, S., and group developed the similar convergent synthetic strategy to MK-6 as well (vitamin K_2 variant) in all the trans forms of side chain through "1+5 convergent synthetic approach" of pentaprenyl chloride with monoprenyl menadione derivative [53]. During this survey, authors found that bromo based polyprenyl substrate was produced $S_n^{2'}$ side reaction product along with the main product. In addition, during $LiEt_3BH$ mediated desulfonation, it evolves a by-product Et_3B and that leads to give cyclized side product. These limitations in the synthetic strategy interrupt the purification processes and effect the yield of the final product MK-6. To minimize these limitations, the reaction strategy was incorporated less efficient leaving group/ more efficient nucleophile (Cl) in the polyprenyl derivative to eliminate all $S_n^{2'}$ side reactions and thus enhance the yield and purity of the product. Further in the reaction sequence, to prevent the side reactions due to the Et_3B by product (from the desulfonation reaction), the corresponding desulfonation reaction was treated with acetic acid to quench the by-product Et_3B. Authors were applied the same strategy to the one-pot synthetic protocol to MK-9 (vitamin K_2 variant). The reaction of phenyl sulfone derivative of monoprenyl menadiol with pentaprenyl chloride was occurred in presence of base (produced by a sequential reaction from the combination of geraniol and farnesol), and it was followed by the desulfonation reaction of the resulting product with $LiEt_3BH$ in presence of $Pd(dppe)Cl_2$. Finally, the product in the desulfonation process was subjected for the oxidation with CAN reagent and it produced the final product MK-6 (**Figure 19**).

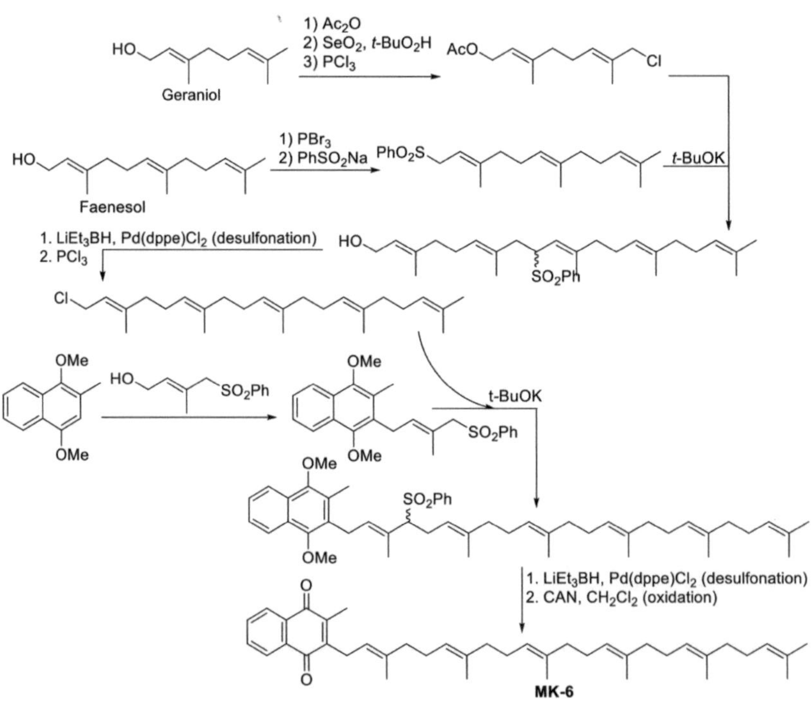

Figure 19.
Convergent synthesis of MK-6.

Lipshutz and group developed a method to introduce a one carbon handle at the C-3 position of menadione molecules and it offered a good synthetic protocol to a wide range of MK-derivatives through a highly probable S_N^2 substitutions and organometallic cross-coupling reactions [55]. The protocol was initiated by the reaction of menadione molecule with formaldehyde and HCl gas, resulting to produce 3-chloromethyl menadione in appreciable yield. The synthesis of vitamin K1 was achieved through the reaction of 3-chloromethyl menadione with phytylalane in presence of $NiCl_2$, Ph_3P and *n*-BuLi. Whereas MK-3 was prepared in the reaction with farnesylalane with *E*-configuration at the coupling position (α-isoprene double bond). The required stereoselective organoalanes [56] were being synthesized by Negishi cross coupling reaction (**Figure 20**).

In the early of 1900s, Saa and group successfully established a protocol by which aldehydes were being used as electrophiles to launch the prenyl side chains in to menadione molecules in the production of MK-2 and MK-4 [57]. During this protocol, bismethyl ether of 2-bromomenadiol undergoes reaction with *n*-BuLi, and generates organolithium reagent through the lithium-bromide ion exchange process. This organolithium reagent attacks on commercially available prenyl aldehydes (citral & geranylgeranial) under BIHY reaction conditions (Birch hydrolysis) produces corresponding benzyl alcohols respectively [58]. The formed alcohol was protected with TMSCl in the presence of HMDS/HMDSZ (hexamethyldisilazane reagent). This protected form of alcohol converted to free methylene group when it reduced in presence of Li & liquid ammonia [59]. The resulting stereochemistry of the α-isoprene double bonds were with more than 95% *E* alkene as presents in

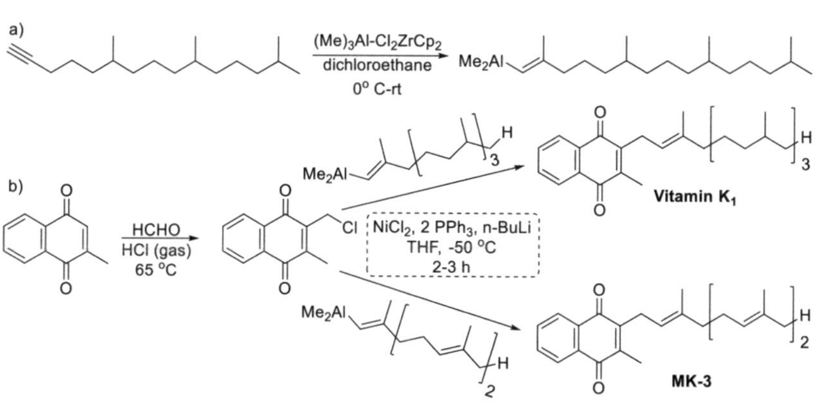

Figure 20.
Synthesis of MK-3 & vitamin K_1 through using phytyl and prenyl organoalane compounds.

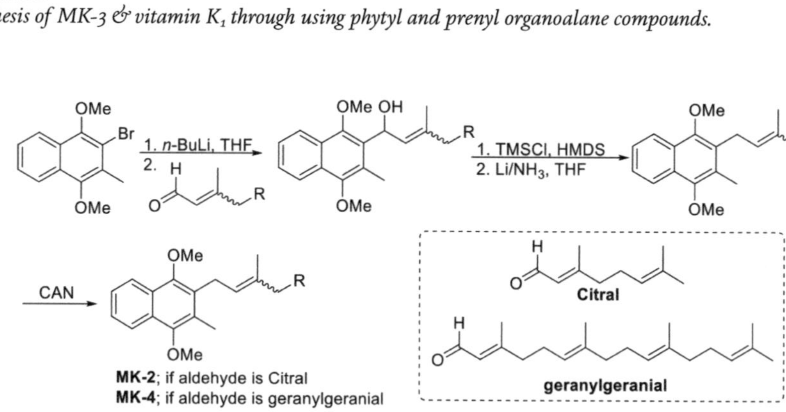

Figure 21.
Synthesis of MK-2 & MK-4 by using birch hydrolysis conditions.

the precursor aldehyde. The methyl ethers of menadione nucleus were removed by CAN to generate MK-2 and MK-4 (**Figure 21**).

4. Brief discussion of vitamin K biology

Biologically these molecules are very important and have been reported for several biomedical purposes. Owing to their severe liphophilicity/ hydrophobicity due to containing multiple isoprene units in the side chain, they are with less solubility and thus causes to difficulties to assess in-vitro studies as these are performed in aqueous solutions. These napthoquinones have been displayed promising biological activities against tubercular [60, 61], cancer [62–64], cardiovascular [65, 66] and diabetes [67, 68].

In mycobacteria, MenJ (*Rv0561c*) is a highly conserved reductase enzyme demonstrated in mycobacteria knock out strains, and that reduces isoprene units in menaquinones through electron transport, thereby the menaquinones found in mycobacteria are in reduced form (**Figure 22**) [3, 69, 70]. But the ATP levels depends on the total menaquinones present in membrane of bacteria and that facilitated by disruption of MenJ functioning of menaquinone reduction. However, these partially saturated menaquinone's role is unclear in various organisms, but they are found to essential for the growth and survival of pathogenic. The mycobacterial electron transport system is increased by hydrogenation of MK-9, so the survival of the bacteria inside J774A.1 macrophage like cells significantly declined by the deletion of MenJ and that found to be not required for bacterial growth in culture. Thereby, MenJ perhaps identified as a conditional drug target for *Mycobacterium tuberculosis* while in the context of infected macrophage [69].

As a known fact that vitamin K is an essential nutrient that displays potential anticancer properties on a variety of tumor cells [71, 72]. Quinones are the important natural and synthetic molecules as they have considerable biological potential. These compounds are display antitumor activity through several mechanism of action. Generally, these molecules have problems with respect to solubility, stability and toxicity. Owing to this reason, these molecules are using as drugs through alternative procedures like controlled-release system of these quinones, and it could be a strategy for improving the pharmacological profile of this class of compounds. Vitamin K mediated mechanisms proceeds to prevent the cell proliferation and growth although unclear, but mostly through oxidative effect and direct arylation of thiols may deplete glutathione and cell cycle arrest. The quinone structure in vitamin k is responsible for the modulation of redox-balance and induction of oxidative stress in cancer cells. The anticancer properties of vitamin K_1 and vitamin K_2 mostly mediated by non-oxidative mechanisms, probably through transcription

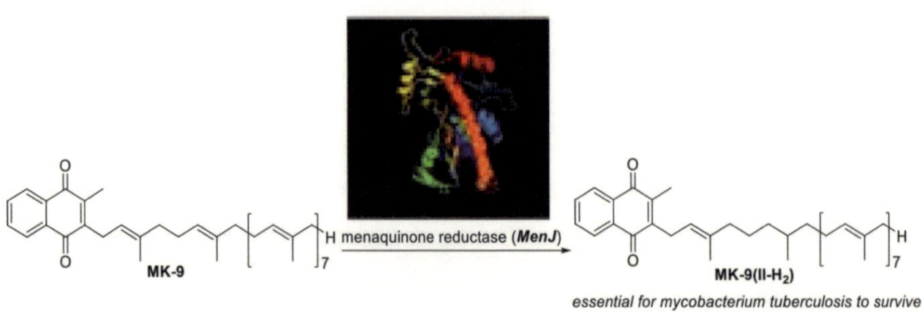

Figure 22.
MenJ reductase reduction transformation of MK-9.

factors, but vitamin K_3 does by reducing oxidative stress and arylation at higher concentrations. It's been evidenced that, bulk doses of vitamin K_2 (2.5 grams given per day) could be safe and not caused to enhancing toxicity levels [73]. Vitamin K_2 also prevents hepatocarcinogenesis in patients with hepatic cirrhosis [74]. Quinones generally undergo one-electron and two-electron reductions, leads to produce semiquinone radicals, as well as hydroquinone's respectively. These factors reduce oxidative stress through the consumption of superoxide radicals and cause to cancer cell homeostasis.

Vitamin K known to reduce complications and improve clinical issues of pre-diabetes and diabetes. Type-2 diabetes mellitus (T2DM) demonstrates when pancreatic β-cells fail to compensate for enduring elevated blood glucose (hyperglycemia) that occurs when glucose uptake in the insulin-sensitive tissues become imbalanced during insulin resistance [75]. Recent clinical trials display T2DM risk reduction was happened with vitamin K supplementation, in addition vitamin K_2 improvs insulin sensitivity through the involvement of vitamin K-dependent-protein osteocalcin metabolism and that favors β-cell proliferation, insulin secretion and sensitivity. Vitamin K_2 shows better effect than vitamin K_1 in the context of T2DM [76].

Vitamin K_2 also useful to the bone and cardiovascular related problems. As we know, lower intake of calcium can decrease the bone mineral density, and thus can increase the risk of bone fractures. Although supplemental calcium helps to enhance bone mineral density and strength (prevent osteoporosis), recent evidences informed that higher consumption of calcium supplements may lead to the risk for heart diseases and also can cause to accelerated deposit of calcium in blood-vessel walls and soft tissues. While the vitamin K_2 is related with the inhibition of arterial calcification and arterial stiffening, which means that increased vitamin K_2 intake could be lower the risk of vascular damage as it activates matrix GLA protein (MGP), and that inhibits the deposits of calcium on the walls, and thus reducing the health risks that are associated with calcium levels [77, 78]. The essential component to the synthesis of Gla-protein family is vitamin K and that is very important to the hemostasis as its deficiency causes to acute and dangerous condition due to excessive bleeding.

5. Conclusion

This chapter summarized various synthetic approaches of different variants of vitamin K and their biological application. Although several methods are available, but menadione and its substrate auxiliaries (for example, Diels-Alder adducts, organometallic compounds and others) are the choices to the synthesis of vitamin K's. Their synthesis proceeds through several reactions like, Friedel-Craft alkylation's, condensation reactions, Claisen rearrangement, Diels-Alder reactions and others and by the involvement of nucleophilic and free radical reactions. It also included the information regarding their natural sources. As the huge importance of vitamin K to the mankind, it is being used as food supplementation because it's not produced by mankind. The major dietary source of vitamin K is phylloquinone, which is synthesized by plants and algae. Vitamin K_2 (various forms of menaquinone; MK-4 to MK14), produces from bacteria in the human gut and plays a lesser role in the provision of vitamin K, since it is taken up by the body to only a limited extent. In infants the development of vitamin K is very low, due to its deficiency they are offering vitamin K immediately after the birth and the initial days life. Also incorporated the utility of vitamin K, as its great role in the blood coagulation, in the maintenance of bone health and healthy nervous system, prevention of cardio vascular disease and diabetes.

Menadione, a synthetic product and being used as a pharmaceutical interested molecule. Menaquinone-4 mainly resides in the brain tissues and generates from a tissue-specific transformation of vitamin K. The metabolism of vitamin K is an essential factor to study further vitamin K biology. Further knowledge in this context of vitamin K proved to be beneficial in many areas of science for example like medicine [79].

Acknowledgements

Uka Tarsadia University, Bardoli, Gujarat, India.

Conflict of interest

The authors declare no conflict of interest.

Author details

Satyanarayana Battula
Tarsadia Institute of Chemical Sciences, Uka Tarsadia University, Bardoli, Gujarat, India

*Address all correspondence to: satyamssd@gmail.com

References

[1] Kroppenstedt RM, Mannheim W. Lipoquinones in Members of the Family Pasteurellaceae. Int. J. Syst. Bacteriol. 1989;39(3):304-308. DOI: 10.1099/00207713-39-3-304.

[2] Collins MD, Jones D. Distribution of isoprenoid quinone structural types in bacteria and their taxonomic implication. Microbiol. Rev. 1981;45(2):316-354.

[3] Upadhyay A, Fontes FL, Gonzalez-Juarrero M, et al. Partial Saturation of Menaquinone in *Mycobacterium tuberculosis*: Function and Essentiality of a Novel Reductase, MenJ. ACS Cent. Sci. 2015;1(6):292-302. DOI: 10.1021/acscentsci.5b00212.

[4] VilchÃ¨ze C, Hartman T, Weinrick B, et al. Enhanced respiration prevents drug tolerance and drug resistance in *Mycobacterium tuberculosis*. Proc. Natl. Acad. Sci. 2017;114(17):4495-4500. DOI: 10.1073/pnas.1704376114.

[5] Nakagawa K, Hirota Y, Sawada N, et al. Identification of UBIAD1 as a novel human menaquinone-4 biosynthetic enzyme. Nature. 2010;468(7320):117-121. DOI: 10.1038/nature09464.

[6] Dunphy PJ, Brodie AF. The structure and function of quinones in respiratory metabolism. In Methods Enzymology; Elsevier: Amsterdam, The Netherlands, 1971; 18:407-461.

[7] Koehn JT, Magallanes ES, Peters BJ, et al. A Synthetic Isoprenoid Lipoquinone, Menaquinone-2, Adopts a Folded Conformation in Solution and at a Model Membrane Interface. J. Org. Chem. 2018;83(1):275-288.

[8] Puffal J, Mayfield JA, Moody DB, Morita YS. Demethylmenaquinone Methyl Transferase Is a Membrane Domain-Associated Protein Essential for Menaquinone Homeostasis in *Mycobacterium smegmatis*. Front. Microbiol. 2018;9(3145). DOI: 10.3389/fmicb.2018.03145.

[9] Conly JM, Stein K. The production of menaquinones (vitamin K2) by intestinal bacteria and their role in maintaining coagulation homeostasis. Prog Food Nutr Sci. 1992;16(4):307-43.

[10] Bentley R, Meganathan R. Biosynthesis of vitamin K (menaquinone) in bacteria. Microbiol. Rev. 1982;46(3): 241-280.

[11] Suvarna K, Stevenson D, Meganathan R, Hudspeth MES. Menaquinone (Vitamin K2) Biosynthesis: Localization and Characterization of the menA Gene from *Escherichia coli*. J. Bacteriol. 1998;180(10):2782-2787. DOI: 10.1128/JB.180.10.2782-2787.1998.

[12] Meganathan R, Kwon. 2009. Biosynthesis of Menaquinone (Vitamin K2) and Ubiquinone (Coenzyme Q). EcoSal Plus 3(2):DOI:10.1128/ecosalplus.3.6.3.3.

[13] Collins MD, Jones D. Distribution of isoprenoid quinone structural types in bacteria and their taxonomic implication. Microbiol. Rev. 1981; 45:316-354.

[14] Cox GB, Gibson F. Biosynthesis of vitamin K and ubiquinone. Relation to the shikimic acid pathway in *Escherichia coli*. Biochim. Biophys. Acta. 1964; 93: 204-206.

[15] Campbell IM, Coscia CJ, Kelsey M, Bentley R. Origin of the aromatic nucleus in bacterial menaquinones. Biochem. Biophys. Res. Commun. 1967; 28:25-29.

[16] Campbell IM. The roles of alanine, aspartate and glutamate in lawsone biosynthesis in *Impatiens balsamina*. Tetrahedron Lett. 1969; 1969:4777-4780.

[17] Robins DJ, Bentley R. Biosynthesis of bacterial menaquinones-evidence for

the involvement of 2-oxoglutaric acid. J. Chem. Soc. Chem. Commun. 1972; 1972:232-233.

[18] Kolappan S, Zwahlen J, Zhou R, Truglio JJ, Tonge PJ, Kisker C. Lysine 190 is the catalytic base in MenF, the menaquinone-specific isochorismate synthase from *Escherichia coli*: Implications for an enzyme family. Biochemistry. 2007; 46:946-953.

[19] Meganathan R. Enzymes from *Escherichia coli* synthesize o-succinyl-benzoic acid, an intermediate in menaquinone (vitamin K2) bio-synthesis. J. Biol. Chem. 1981; 256: 9386-9388.

[20] Palaniappan C, Sharma V, Hudspeth ME, Meganathan R. Menaquinone (vitamin K2) biosynthesis: evidence that the *Escherichia coli* menD gene encodes both 2-succinyl-6-hydroxy-2,4-cyclohexadiene-1-carboxylic acid synthase and alpha-ketoglutarate decarboxylase activities. J. Bacteriol. 1992; 174:8111-8118.

[21] Jiang M, Cao Y, Guo ZF, Chen MJ, Chen XL, Guo ZH. Menaquinone biosynthesis in *Escherichia coli*: Identification of 2-succinyl-5-enolpyruvyl-6-hydroxy-3-cyclohexene-l-carboxylate as a novel intermediate and re-evaluation of MenD activity. Biochemistry. 2007; 46:10979-10989.

[22] Sieminska EAL, Macova A, Palmer DRJ, Sanders DAR. Crystallization and preliminary X-ray analysis of (1R, 6R)-2-succinyl-6-hydroxy-2,4-cyclohexadiene-1-carboxylate (SHCHC) synthase (MenD) from *Escherichia coli*. Acta Crystallogr. Sec. F Strut. Biol. Crystal Commun. 2005; 61:489-492.

[23] Jiang M, Cao Y, Guo ZF, Chen MJ, Chen XL, Guo ZH. Menaquinone biosynthesis in *Escherichia coli*: Identification of 2-succinyl-5-enolpyruvyl-6-hydroxy-3-cyclohexene-l-carboxylate as a novel intermediate and re-evaluation of MenD activity. Biochemistry. 2007; 46:10979-10989.

[24] Meganathan R, Bentley R. Thiamine pyrophosphate requirement for o-succinylbenzoic acid synthesis in *Escherichia coli* and evidence for an intermediate. J. Bacteriol. 1983; 153: 739-746.

[25] Bryant RW Jr, Bentley R. Menaquinone biosynthesis: conversion of o-succinylbenzoic acid to 1,4-dihydroxy-2-naphthoic acid and menaquinones by *Escherichia coli* extracts. Biochemistry. 1976;15:4792-4796.

[26] Shaw DJ, Guest JR, Meganathan R, Bentley R. Characterization of *Escherichia coli* men mutants defective in conversion of o-succinylbenzoate to 1,4-dihydroxy-2-naphthoate. J. Bacteriol. 1982; 152:1132-1137.

[27] Bentley, R. Molecular asymmetry in biology. Vol. 2. New York: Academic press, Inc.; 1970. p.480-486.

[28] Baldwin RM, Snyder CD, Rapoport H. Biosynthesis of bacterial menaquinones. Dissymmetry in the naphthalenic intermediate. Biochemistry. 1974; 13:1523-1530.

[29] Suzuki K, Ueda M, Yuasa M, Nakagawa T, Kawamukai M, Matsuda H. Evidence that *Escherichia coli* ubiA product is a functional homolog of yeast COQ2, and the regulation of ubiA gene expression. Biosci. Biotechnol Biochem. 1994; 58:1814-1819.

[30] Shineberg B, Young IG. Biosynthesis of bacterial menaquinones: the membrane-associated 1,4-dihydroxy-2-naphthoate octaprenyltransferase of *Escherichia coli*. Biochemistry. 1976; 15:2754-2758.

[31] Jackman LM, O'Brien IG, Cox GB, Gibson F. Methionine as the source of methyl groups for ubiquinone and vitamin K: a study using nuclear magnetic

resonance and mass spectrometry. Biochim. Biophys. Acta. 1967; 141:1-7.

[32] Alison MD, Richard JP, Mark EH, Andrew DA. The Synthesis of Naturally Occurring Vitamin K and Vitamin K Analogues. Curr. Org. Chem. 2003; 7(16):1625-1634.

[33] Fieser LF, Campbell WP, Fry EM, Gates MD. Synthetic approach to vitamin K 1. J. Am. Chem. Soc. 1939; 61(9):2559-2559.

[34] Binkley SB, Cheney LC, Holcomb WF, et al. The constitution and synthesis of vitamin K1. J. Am. Chem. Soc. 1939;61(9):2558-2559.

[35] Almquist HJ, Klose AA. Synthetic and natural antihemorrhagic compounds. J. Am. Chem. Soc. 1939;61(9):2557-2558.

[36] Lindlar, H. Patent No. 320582, Switzerland, 1953.

[37] Sato K, Inoue S, Saito K. A new synthesis of vitamin K via π -allyinickel intermediates. j. chem. soc., perkin trans. 1. 1973(0):2289-2293.

[38] Snyder CD, Rapoport H. Synthesis of menaquinones. J. Am. Chem. Soc. 1974;96(26):8046-8054.

[39] Evans DA, Hoffman JM. Regiospecific quinone isoprenylation. Examples of remarkably facile [3,3] sigmatropic processes. J. Am. Chem. Soc. 1976;98(7):1983-1984.

[40] Chenard BL, Manning MJ, Raynolds PW, Swenton JS. Organocopper chemistry of quinone bisketals. Application to the synthesis of isoprenoid quinone systems. J. Org. Chem. 1980;45(3):378-384.

[41] Syper L, Kloc K, Mz.xl, lochowski J. Synthesis of ubiquinone and menaquinone analogues by oxidative demethylation of alkenylhydroquinone ethers with argentic oxide or ceric ammonium nitrat. Tetrahedron. 1980;36(1):123-129.

[42] Maruyama K, Imahori H. Allylation of Quinones via Photoinduced Electron-Transfer Reactions from Allylstannanes. Bull. Chem. Soc. Jpn. 1989;62(3): 816-825.

[43] Liebeskind LS, Foster BS. Stannylquinones. Synthesis and utilization as quinone carbanion synthetic equivalents. J. Am. Chem. Soc. 1990;112(23):8612-8613.

[44] Tso H-H, Chen Y-J. A convenient one-flask synthesis of vitamin K. J. Chem. Res. Synop. 1995;3:104-105.

[45] Mal D, Ghosh K, Jana S. Synthesis of Vitamin K and Related Naphtho-quinones via Demethoxycarbonylative Annulations and a Retro-Wittig Rearrangement. Org. Lett. 2015;17(23): 5800-5803.

[46] Ryall RP, Nandi DL, Silverman RB. Substituted vitamin K epoxide analogs. New competitive inhibitors and substrates of vitamin K1 epoxide reductase. Journal of Medicinal Chemistry. 1990;33(6):1790-1797.

[47] Shearer, M. J. In Encyclopaedia of Food Science, Food Technology and Nutrition; Macrae, R., Robinson, R. K., Sadler, M. J., Eds.; Academic Press: London, 1993, p 4804.

[48] Hirschmann R, Miller R, Wendler NL. The Synthesis of Vitamin K1. J. Am. Chem. Soc. 1954, 76, 4592-4594.

[49] Schmida R, Antoulas So, RÃ ttimann A, Schmid M, Vecchi M, Weiserb H. Synthesis of All Four Stereoisomers of (E)-Vitamin KT (Phylloquinone), Analysis of Their Diastereoisomeric and Enantiomeric Purities and Determination of Their Biopotencies. Helv. Chim. Acta 1990;73(5):1276-1299.

[50] Jacobsen N, Torssell K. Radikalische Alkylierung von Chinonen: Erzeugung

von Radikalen in Redoxreaktionen. Liebigs Ann. Chem. 1972;763(1):135-147.

[51] Yamago S, Hashidume M, Yoshida J-i. A new synthetic route to substituted quinones by radical-mediated coupling of organotellurium compounds with quinones. Tetrahedron. 2002;58(34): 6805-6813.

[52] Rüttimann, A. Recent Advances in the Synthesis of K-Vitamins. Chimia 1986, 40, 290-306.

[53] Yerramsetti N, Dampanaboina L, Mendu V, Battula S. Synergistic factors ensue high expediency in the synthesis of menaquinone [K2] analogue MK-6: Application to access an efficient one-pot protocol to MK-9. Tetrahedron.76(49):131696.

[54] Baj A, WaÅ,ejko P, Kutner A, Kaczmarek Åu, Morzycki JW, Witkowski Sa. Convergent Synthesis of Menaquinone-7 (MK-7). Org. Process Res. Dev. 2016;20(6):1026-1033.

[55] Lipshutz BH, Kim SK, Mollard P. An Expeditious Route to COQn, Vitamins K1 and K2, and Related Allylated para-Quinones Utilizing Ni(0) Catalysis. Tetrahedron 1998;54:1241-1253.

[56] Negishi E, Van Horn DE, Yoshida T. Controlled carbometalation. 20. Carbometalation reaction of alkynes with organoalene-zirconocene derivatives as a route to stereo-and regiodefined trisubstituted alkenes. J. Am. Chem. Soc. 1985; 107:6639-6647.

[57] Garcias X, Ballester P, Capo M, Saá JM. 2.DELTA.-Stereocontrolled Entry to (E)- or (Z)-Prenyl Aromatics and Quinones. Synthesis of Menaquinone-4. J. Org. Chem. 1994; 59:5093-5096.

[58] Ballester P, Capo M, Garcias X, Saá JM. Synthesis of prenylated quinones by the oxidative degradation approach. Birch vs vinylogous Birch hydrogenolysis (BIHY vs VIBIHY) in controlling 2.DELTA. stereochemistry of the prenyl chain. J. Org. Chem. 1993; 58:328-334.

[59] Ballester, P.; Saá, J.; Capo, M. Prenylation of Aromatics. Lithium-Ammonia Reduction of 0-Silylated Tertiary Cinnamyl Alcohols. Tetrehedron Lett. 1990, 31, 1339-1342.

[60] Gengenbacher M, Kaufmann SHE. Mycobacterium tuberculosis: success through dormancy. FEMS Microbiol. Rev. 2012;36(3):514-532. DOI: 10.1111/j.1574-6976.2012.00331.x.

[61] Bahuguna A, Rawat DS. An overview of new antitubercular drugs, drug candidates, and their targets. Med. Res. Rev. 2020;40(1):263-292. DOI: 10.1002/med.21602.

[62] Vitor Francisco F, Caroline Deckmann N, Patricia Garcia F, Debor Omena F, Fernando de Carvalho da S. Strategies for Increasing the Solubility and Bioavailability of Anticancer Compounds: β-Lapachone and Other Naphthoquinones. Curr. Pharm. Des. 2016;22(39):5899-5914. DOI: 10.2174/1381612822666160611012532.

[63] Setoguchi S, Watase D, Matsunaga K, et al. Enhanced antitumor effects of novel intracellular delivery of an active form of menaquinone-4, menahydroquinone-4, into hepatocellular carcinoma. Cancer Prev Res (Phila). 2015;8(2):129-138. DOI: 10.1158/1940-6207.CAPR-14-0292.

[64] Dasari S, Ali SM, Zheng G, et al. Vitamin K and its analogs: Potential avenues for prostate cancer management. Oncotarget. 2021/01/29 2017;5(0):57782-57799. DOI: 10.18632/oncotarget.17997.

[65] Ingles DP, Cruz Rodriguez JB, Garcia H. Supplemental Vitamins and Minerals for Cardiovascular Disease Prevention and Treatment. Curr Cardiol Rep. 2020;22(4):22. DOI: 10.1007/s11886-020-1270-1.

[66] Maresz K. Proper Calcium Use: Vitamin K2 as a Promoter of Bone and Cardiovascular Health. Integra Med (Encinitas, Calif.). 2015;14(1):34-39.

[67] Karamzad N, Maleki V, Carson-Chahhoud K, Azizi S, Sahebkar A, Gargari BP. A systematic review on the mechanisms of vitamin K effects on the complications of diabetes and pre-diabetes. BioFactors. 2020;46(1):21-37. DOI: 10.1002/biof.1569.

[68] Li Y, Chen JP, Duan L, Li S. Effect of vitamin K2 on type 2 diabetes mellitus: A review. Diabetes Res Clin Pract. 2018;136:39-51. DOI: 10.1016/j.diabres.2017.11.020.

[69] Kumar S, Koehn JT, Gonzalez-Juarrero M, Crans DC, Crick DC. Mycobacterium tuberculosis Survival in J774A.1 Cells Is Dependent on MenJ Moonlighting Activity, Not Its Enzymatic Activity. ACS Infect. Dis. 2020;6(10):2661-2671. DOI: 10.1021/acsinfecdis.0c00312.

[70] Upadhyay A, Kumar S, Rooker SA, et al. Mycobacterial MenJ: An Oxidoreductase Involved in Mena-quinone Biosynthesis. ACS Chem. Biol. 2018;13(9):2498-2507. DOI: 10.1021/acschembio.8b00402.

[71] Wang Z, Wang M, Finn F, Carr BI. The growth inhibitory effects of vitamins K and their actions on gene expression. Hepatology. 1995;22(3): 876-882.

[72] Sata N, Klonowski-Stumpe H, Han B, Häussinger D, Niederau C. Menadione Induces Both Necrosis and Apoptosis in Rat Pancreatic Acinar AR4-2J Cells. Free Radical Biology and Medicine. 1997;23(6):844-850. DOI: 10.1016/S0891-5849(97)00064-6.

[73] Margolin KA, Akman SA, Leong LA, et al. Phase I study of mitomycin C and menadione in advanced solid tumors. Cancer Chemother Pharmacol. 1995;36(4):293-298. DOI: 10.1007/BF00689046.

[74] Tamori A, Habu D, Shiomi S, Kubo S, Nishiguchi S. Potential role of vitamin K(2) as a chemopreventive agent against hepatocellular carcinoma. Hepatol Res. 2007;37(Suppl 2):S303-307.

[75] Karamzad N, Maleki V, Carson-Chahhoud K, Azizi S, Sahebkar A, Gargari BP. A systematic review on the mechanisms of vitamin K effects on the complications of diabetes and pre-diabetes. BioFactors 2019; 46(1):21-37.

[76] Aguayo-Ruiz JI, GarcÃa-CobiÃ¡n TA, Pascoe-GonzÃ¡lez S, SÃ¡nchez-EnrÃquez S, Llamas-Covarrubias IM, GarcÃa-Iglesias T, LÃ³pez-Quintero A, Llamas-Covarrubias MA, Trujillo-Quiroz J, Rivera-Leon EA. Effect of supplementation with vitamins D3 and K2 on undercarboxylated osteocalcin and insulin serum levels in patients with type 2 diabetes mellitus: a randomized, double-blind, clinical trial. Diabetol Metab Syndr 2017; 12(1):73.

[77] Maresz K. Proper Calcium Use: Vitamin K2 as a Promoter of Bone and Cardiovascular Health. Integr Med (Encinitas). 2015;14(1):34-39.

[78] van Ballegooijen AJ, Beulens JW. The Role of Vitamin K Status in Cardiovascular Health: Evidence from Observational and Clinical Studies. Curr Nutr Rep. 2017;6(3):197-205.

[79] Card DJ, Gorska R, Cutler J, Harrington DJ. Vitamin K metabolism: current knowledge and future research. Mol Nutr Food Res. 2014 58(8): 1590-1600.

Chapter 3

Physiological and Cellular Functions of Vitamin K on Cardiovascular Function

Meneerah A. Aljafary, Hussah Alshwyeh, Nada Alahmadi, Adeeb Shehzad, Huseyin Tombuloglu, Zagit Gaymalov, Abdelqader Homieda and Ebtesam Al-Suhaimi

Abstract

This chapter reviews the physiological and cellular functions of vitamin K in the cardiovascular system based on the latest pre-clinical and clinical evidence. Vitamin K belongs to a family of structurally similar fat-soluble vitamins, actively required by the body for the synthesis of essential proteins as well as regulate blood clotting, bone metabolism and calcium level. The authors emphasize the quintessential association between dietary vitamin K2 and cardiovascular diseases shown in various studies. The association, through the vitamin K - dependent hormones, plays a primary role in regulating calcification of different cell types, especially their role in calcification of the vascular endothelial cells. The consequences of vitamin K deficiency in the vascular system are unfavorable, shown in various clinical studies on statins - well-known inhibitors of vitamin K production in the body. New clinical insights suggest that vitamin K levels in the body and its dietary supplementation play a crucial role in cardiovascular disease prevention. There is negative influence of these antagonist's pate in vascular composition and functions. Therefore, there is a need for prospective studies to make more in-depth exploration and increase the current understanding of this critical relationship to confidently apply such knowledge to prevent cardiovascular diseases and improve their outcomes.

Keywords: Vitamin K, hormone, heart, cardiac disease, vascular system, gene expression, statin

1. Introduction

Vitamin K applies to fat-soluble vitamins, which are similar in structure and essential in the blood coagulation process and to control the calcium mineral binding in bones and other tissues. The discovery of vitamin K can be attributed to the observations of a high incidence of bleeding in chickens on a low-lipid diet during the 1930s [1]. Until the 1970s, it was believed that vitamin K was essential exclusively for homeostasis, maintaining an adequate blood supply, and preserving vascular integrity in animals and humans. Today, we know that vitamin K is involved in gamma-carboxylation as a co-factor in several essential proteins located within the bone, heart, and blood vessels. Moreover, only in the presence of vitamin K, specific essential proteins called vitamin K-dependent proteins (VKDPs) are able to switch from inactive uncarboxylated forms to active carboxylated forms. Vitamin K allows switching VKDPs to their active states by the carboxylation of VKDPs glutamic acid (Glu) residues in specific tissues and organs [2]. VKDPs include seven proteins involved in blood coagulation (coagulation factor II, VII, IX, and X, and anticoagulant proteins C, S, and Z); proteins responsible for bone mineralization (osteocalcin (OC) and matrix gamma-carboxyglutamic acid (Gla)-protein (MGP)); and recently discovered proteins, including growth arrest-specific gene 6 (Gas-6), the transmembrane Gla proteins (TMG3 and TMG4), the proline-rich Gla proteins (PRGP1 and PRGP2), the Gla-rich protein (GRP), periostin and transthyretin [3–7]. In nature, there are two main variants of vitamin K: phylloquinone (or vitamin K1) can be found in some green vegetables, and menaquinone (or vitamin K2) can be found in some meat, fermented milk, and fermented soybean products. Structurally, vitamers differ in their degree of saturation and side-chain lengths; vitamin K2 is more biologically active than vitamin K1 and circulates longer in the body [8]. Cardiovascular diseases (CVDs) comprise a group of disorders affecting the heart and blood vessels that cause conditions such as coronary heart disease (CHD) and cerebrovascular disease, which affect the blood vessels that supply the brain [9]. In recent years, numerous physiological studies have pointed out the critical role of vitamin K as an anti-vascular calcification (VC) factor. VC is recognized as an autonomous and prominent risk factor for CVDs, and several human observational studies have shown a positive correlation between low vitamin K supplementation and VC [10, 11]. Meta-analysis studies showed that patients with regular dietary of vitamin K showed significantly less VC while maintaining vascular stiffness compared with patients with no vitamin K supplementation [12]. In this chapter, we will cover the following topics: the importance of dietary vitamin K; physiological functions of vitamin K beyond blood coagulation; vitamin K-dependent hormones; physiological and protective roles of vitamin K on cardiovascular (CV) processes; cellular and molecular mechanisms of vitamin K in the vascular cells and whether vitamin K promotes or encounters statins.

2. The importance of dietary vitamin K in CVD

Vitamin K is vital for healthy bones and the heart and increases blood clotting. Although a deficiency of vitamin K is not common, its deficiency may affect the body over time. Bleeding and weak bones, as well as higher CV risks, are some consequences of vitamin K deficiency [13, 14]. Hence, vitamin K intake must not be ignored. Usually, the daily value (DV) of 120 mcg of vitamin K is sufficient in adult males and less than this in females and children (**Table 1**) [17, 18]. Vitamin K can be in both forms in our diets, where phylloquinone can be sourced from leafy green vegetables, and menaquinone can be sourced from animal-based food that includes meats, fermented dairy, fermented soybeans, and dietary supplements [19]. Vitamin K2 is also available on the market in the form of synthetic menaquinone-4 (MK-4) and menaquinone-7 (MK-7) in natural or synthesized form. Naturally, in animals, MK-4 is more commonly produced, while MK-7 (as well as MK-5 to MK-14) is made by bacteria. Many animals can convert vitamin K1 to vitamin K2 (MK-4). Vascular health can be improved by ensuring the consumption of sufficient amounts of vitamin K2. Vitamin K2 stimulates MGP, which prevents calcium from depositing inside the vessel walls. When calcium is not deposited in the arteries, it offers dual benefits of clear arteries as well as the availability of calcium for various functions in the human body [20]. Presently, MGP is found to be highly effective for the modulation of arterial calcification. Although MGP binds

Type of product	Amount of Vitamin K2 (mcg/100 g)	% of daily value (120mcg)	Ref.
Ground beef	9.4	8	[15]
Beef liver	106	88	[15]
Beef kidney	5.7	5	[15]
Chicken	60	50	[15]
Chicken liver	13	11	[15]
Duck breast	5.5	5	[15]
Goose liver	369	308	[15]
Pork chops	69	57	[2]
Pork liver	7.8	7	[2]
Bacon	35	29	[2]
Eggs	˂1	˂1	[16]
Whole milk	9	8	[3]
Fish	˂1	˂1	[3]

Table 1.
Amount (% of daily value) of vitamin K2 in 100 g of animal products that prevent insufficiency in man.

calcium to protect calcification within blood vessels, it needs to be first activated via an adequate dose of vitamin K2 [20]. A total of 4807 healthy individuals from both genders with ages above 55 were involved in a population-based study conducted in Rotterdam. The study aimed to investigate the impact of dietary intake of vitamin K on calcification within the aorta, CVDs, and all-cause mortality [21]. It was found that the risk of calcification within the arteries and CVDs were reduced by half, while the all-cause mortality risk was reduced by one-quarter as a result of a higher dietary intake of vitamin K2 (minimum daily intake of 32 μg) instead of vitamin K1 [22]. Another population-based study that involved 16,000 healthy females with ages between 49 and 70 was conducted; this study showed corresponding results. The study participants were selected from the cohort population of the European Prospective Investigation into Cancer and Nutrition (EPIC) study [23]. The data obtained from the study depicted that vitamin K2 instead of vitamin K1 had to be consumed in high quantities to prevent CV disorders. The data revealed that there was a 9% reduction in the risk of CHDs with every dose of 10 μg of vitamin K2 (taken as MK-7, MK-8, and MK-9). In the Netherlands, ultrasound and pulse wave velocity methods were utilized by the researchers working at the research and development (R&D) Group Vita K of Maastricht University to study 244 healthy postmenopausal females [23]. The subjects were observed for three years. Some of the subjects were administered a dose of vitamin K2 (180 μg) in the form of MK-7 (as MenaQ7 from NattoPharma), while some of the participants were given a placebo capsule every day for three years [22]. This was conducted on a random basis. At the end of the treatment, the group given vitamin K2 supplementation demonstrated a steady decline in stiffness index than the placebo group that showed a slight rise in the index. The study outcomes showed the positive impact of MenaQ7 on vascular health by enhancing vascular elasticity in females with stiff arteries and by suppressing age-related artery-wall stiffening. The researchers also found that the CV conditions improved as the subjects were administered a nutritional dose of vitamin K2 in the form of MK-7 (as MenaQ7). Moreover, if vitamin K2 is taken on a daily basis, there is a high chance of preventing the hardening of arteries [23, 24].

3. Vitamin K-dependent hormones

The growing clinical evidence suggests that regular vitamin K supplementation may improve bone structure, prevent VC, improve the body's sensitivity to the insulin hormone, which increases the life expectancy and treatment outcome in patients [25]. In the past ten years, more evidence has been published supporting the hypothesis that vitamin K2 should be considered a hormone. Vitamin K2 was found to activate many genes directly and indirectly by binding to the intranuclear receptor SXR, activating sirtuins and/or histone deacetylases (HDACs) responsible for cell-type determination and specific cell functions [26]. A study by Lanham et al. on rats and their offspring explored the effect of a high-fat diet on bone development and vascular development, particularly the role of VKDPs, including Gas-6, MGP) and OC [27]. The study also shows the importance of proper nutrition during pregnancy. During the study, the team observed increased levels of Gas-6 proteins, increased expression of the gene responsible for vitamin K-dependent gamma-glutamyl carboxylase (GGCX) in the cardiovascular tissues, while decreased levels of MGP in the femoral bones of female offsprings of high-fat dietary fed mothers [27]. The osteoblastic synthesis gives rise to OC production, deposited into bone or released into circulation, giving the histological measures of bone formation. OC's structure is greatly affected by vitamin K-dependent Gla residues, resulting in bone mineral maturation. The circulating uncarboxylated OC (unOC) levels have been applied as biomarkers for vitamin K deficiency and correlated with age-related bone loss. In animals, in-vivo and in-vitro tests have revealed unOC as an active hormone affecting glucose metabolism; however, the results are inconclusive on human levels and need to be investigated further [28]. Post-translational GGCX enzymes detected both hepatically and extrahepatically are critical for the functionality of Gla residues in VKDPs. OC (bone-derived protein) has been associated with energy metabolism as the skeleton system has been considered an endocrine organ [29]. Via molecular mechanisms, OC mediates vitamin K positive effects, improves insulin resistance, lipid, and glucose profiles. OC is also detected by insulin to regulate bone mineralization. It has been hypothesized that normal VKDP carboxylation is an essential step in the prevention of vascular endothelial calcification [30]. Vitamin K2 has been found to affect bone and CV health. A study of the vitamin K2 homolog MK-7 found that serum levels increased, as evidenced by healthy Japanese women, who supplemented their diet with MK-7, which can be particularly important for extrahepatic tissue health [31]. In a study of the murine model, the importance of the vitamin K-dependent MGP on the inhibition of extraskeletal calcification was suggested. With a high dose dietary supplementation of MK-7, the induced VC was inhibited, and the aortic alkaline phosphatase tissue concentration was reduced [32].

4. Molecular mechanism of vitamin K-dependent calcification on vascular system

Carboxylation is one of the post-translational modifications on proteins and is essential for the activity of VKDPs. The activation of VKDPs, which includes coagulation factors, OC, MGP, Gas-6, GRP, and periostin, is achieved by carboxylation of the proteins' Glu residues [33, 34]. The carboxylation of VKDPs happens in the case of an abundance of vitamin K, which is required for the activation of the GGCX enzyme (**Figure 1**) [35]. This enzyme adds carboxyl groups on Glu residues of VKDPs and converts them to Gla (**Figure 2**) [35]. This conversion enables the Gla residues to

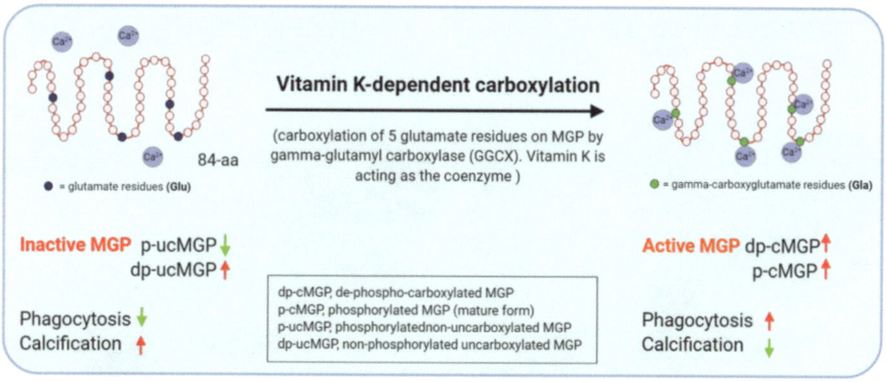

Figure 1.
Vitamin K-dependent post-translational carboxylation of MGP protein.

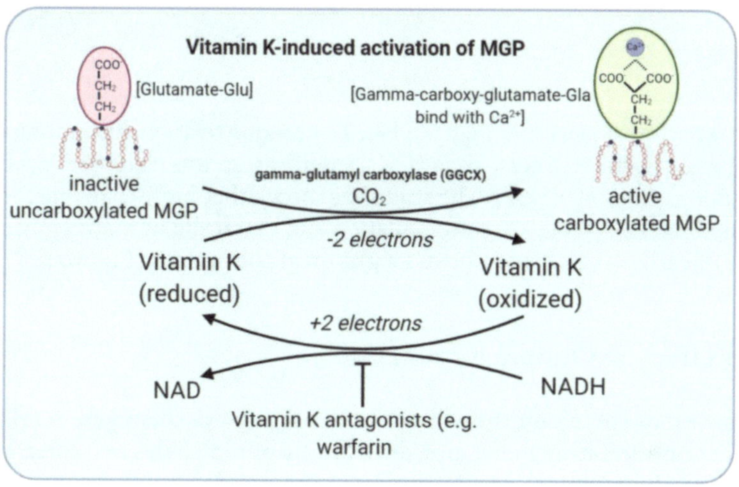

Figure 2.
General overview of vitamin-K induced activation of VKPDs.

capture free Ca^{2+} ions that are circulating in the vascular system [36, 37]. For instance, there are five Glu residues on MGP, which is the only protein known as an inhibitor of arterial calcification. The Glu residues on the protein are carboxylated and converted to Gla, the active form of the protein. Without being activated, MGP is unable to hold free Ca^{2+} ions, which are eventually deposited in the vascular system and cause VC, such as calcium deposits and atherosclerotic plaques (**Figure 3**) [38, 39]. After the inflammation and hyperlipidemia, the soft-tissue mineralization phenomena occur in vascular smooth muscle cells (VSMCs), leading to their hardening and differentiation into osteoblast-type cells [40]. In addition to the mineralization phenomena, it has been hypothesized that active forms of MGP proteins may attach to the calcified crystals in the vasculature resulting in apoptotic bodies and vesicles. Another assumption is their potential to hinder VSMCs' trans-differentiation into an osteogenic phenotype [34, 41, 42]. A three-year clinical study by Shea et al. with 229 patients who were routinely given dietary vitamin K versus 223 patients in the placebo group showed that the addition of dietary vitamin K significantly correlated with decreased levels of calcium in coronary arteries [43]. In addition to carboxylation, post-translational phosphorylation of serine residues occurs on MGP. The enzyme casein kinase adds phosphate groups on three serine residues, regulating the secretion

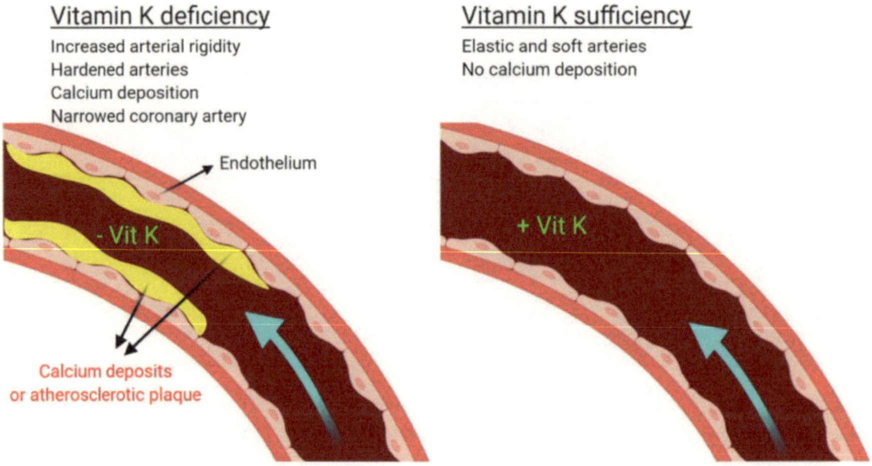

Figure 3.
The consequences of vitamin K deficiency in vascular system.

of protein into the extracellular matrix [44]. The unique relationship among circulating MGP forms, aortic stiffness, and arterial calcification was proposed in a recent article by Roumeliotis et al. [37]. The study has shown that more than one form of the MGP protein can be detected in the circulation and extracellular matrix governed by the degree of carboxylation and phosphorylation of the protein (**Figure 1**) [37].

5. Statins effect on vitamin K2 function

Statins, or more precisely, 3-hydroxy-3-methylglutaryl-coenzyme A (HMG-CoA) reductase-inhibiting molecules, are a family of molecules that interfere with cholesterol synthesis and induce the uptake of the low-density lipoproteins (LDLs) in the body [45]. Available as a prescription since 1987 [46], statins today are one of the most commonly prescribed medications worldwide [47]. However, some researchers are starting to suggest that physicians may be overprescribing statins to their patients [48–50]. A recent study published in the Annals of Internal Medicine journal found that the statins' potential side effects seem to outweigh the benefits for people whose 10-year CVD risk is approximately 7.5–10% [51]. The United States Food and Drug Administration's consumer update from 2017, named "Controlling Cholesterol with Statins," states that statins have been linked to associated muscle symptoms and a high chance of developing type 2 diabetes in patients [52]. It seems crucial to understand the exact phenomena behind statins' mechanism of action and clarify the medical community's created bias [53]. One potential explanation is given by the Kinjo Gakuin University group led by Harumi Okuyama, who suggests that statins may have an essential role in the increased probability of developing diabetes and arteriosclerosis [54] via the inhibition of vitamin K2 synthesis [55]. Researchers have indicated that statins may inhibit vitamin K2 production via the inhibition of geranylgeranyl diphosphate (GGPP) synthesis by HMG-CoA reductases (**Figure 4**) [56]. Some authors hypothesize that prolonged HMG-CoA reductase suppression by chronic statins treatment could adversely affect patients by diminishing the vitamin K2 supply to their bodies [56]. This phenomenon may be an essential factor in diabetes, atherosclerosis, and osteoporosis causation. The recently aggregated data from available randomized controlled trials and observational studies suggest a 10 to 45 percent higher risk of new-onset development of

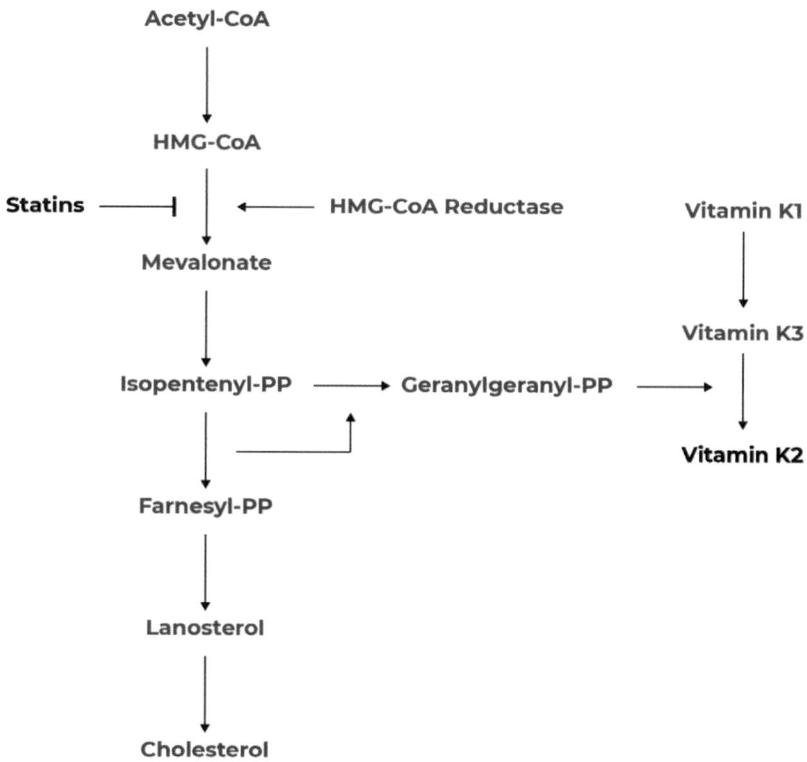

Figure 4.
Statins and vitamin K2 related biochemical pathways. Statins inhibit HMG-CoA reductase in the mevalonate pathway. Geranylgeranyl-PP is essential for the synthesis of vitamin K2 from vitamin K1.

diabetes mellitus type 2 in statin patients than non-users [57]. Studies have shown that postmenopausal women taking statins are 150% more likely to develop type 2 diabetes [58]. In a large clinical study (n = 2,142), Cederberg et al. show a 46% increase in the risk of developing diabetes alongside decreased insulin secretion and overall body sensitivity to insulin [59]. Furthermore, particular varieties of statins—simvastatin and atorvastatin—showed a dose-dependent effect on insulin sensitivity and its secretion in patients [60]. These clinical observations may collectively suggest a relationship between statins' inhibition of Vitamin K2 and GGPP production and statins' influence on insulin synthesis, secretion, and sensitivity. Insufficient levels of vitamin K2 may be linked to atherosclerosis and other CVDs [61]. In the large population-based Rotterdam study (n = 7,983 men and women, age > 55 years), Geleijnse et al. report a positive correlation between the reduced risk of CHD with regular dietary consumption of vitamin K2 [59]. The authors hypothesize the link between the depletion of vitamin K2 levels and severe coronary artery calcification in patients.

Several clinical studies have demonstrated that administering vitamin K2 (but not vitamin K1) was an effective method of osteoporosis fracture prevention in patients [62, 63]. Studies have shown that vitamin K2 up-regulates bone markers' expression and sustains lumbar bone mineral density in patients [63]. Pre-clinical and clinical evidence shows that statins may have an essential role in reducing the supply of vitamin K2 to the tissues through the body. Additional studies are required to explore the mechanisms for statin-associated diseases that were identified in pre-clinical and clinical studies, including diabetes, atherosclerosis, osteoporosis, chronic kidney disease, and cancer, and the effects of the various types of statins on the vitamin K2 synthesis, delivery, and accumulation in the body.

6. Physiological roles of vitamin K on CV functions

There is growing preclinical and clinical evidence of the crucial role of vitamin K in VC prevention [64]. Several VKDPs are regulated by vitamin K share entirely Gla, which is the remarkable amino acid produced by the posttranslational modification of the vitamin K-mediated enzyme GGCX [65]. The prothrombin molecule is carboxylated at ten glutamyl residues to produce the active form of prothrombin. These Gla are deposited at the amino-terminal domain of all VKDP, which share a common amino acid sequence [66]. Studies have shown that Gla also regulates calcium due to the placement of Gla in calcium-binding sites of the protein [67].

Available literature has limitations regarding the average requirement of vitamin K for normal homeostasis. In 2001, the United States Pharmacopeia Health and Medicine Division established adequate intake (AI) values based on median intake values reported by the National Health and Nutrition Examination Survey (NHANES) III; the AI value for vitamin K1 was set to 90 μg/dl for adult females and 120 μg/dl for adult males [67, 68]. This study does not advocate that this concentration of vitamin K will be enough to maintain the carboxylation of VKDPs. Undercarboxylated, biologically inactive Gla proteins are caused by vitamin K deficiency, resulting in the synthesis of calcification, which is considered a risk factor for VC and CVD [67, 68].

CVD is a cluster of abnormal conditions, such as CHD, and influences the functions of the heart and blood vessels that supply blood to various parts of the body [69]. Heart diseases and stroke are the primary causes of death and disability worldwide. According to the American Heart Association 2020 report, the age-adjusted death rate of CVD is 219.4 per 100,000, which means that someone is dying of CVD every 37 seconds, with a total of 2,353 deaths from CVD each day in the U.S. Consistent with these data, there are approximately 795,000 new or recurrent strokes each year, as well as approximately 401 deaths from stroke each day, based on data for previous years [70]. CVD-related diseases, such as angina, carotid artery diseases, and peripheral artery diseases, are characterized by the formation of fatty deposits in the arteries, which is known as atherosclerosis. These deposits consist of calcium, cellular waste products, fatty substances, cholesterol, and fibrin (a clotting material in the blood), which ultimately leads to narrowing and blockage of the arteries and reduced blood flow to the heart muscle. Studies have shown that lifestyle, healthy diets, vitamins, and physical activity may have a potential role in preventing the development of CVD [71, 72].

There is compelling evidence that vitamin K is involved in various biological processes in the body and mediates anti-calcification, anti-cancer, bone-forming, and insulin-sensitization effects, and plays a vital role in the prevention management of CVD (**Figure 5**) [72]. It has been reported that vascular deficiency of vitamin K can lead to CVD by increasing calcium deposition and coronary artery calcification because vitamin K-synthesized osteocalcin and MGP strongly inhibit VC by regulating bone metabolism. Previous studies have shown a strong association between reduced intake of vitamin K and the development of coronary calcification, advocating that adequate vitamin K intakes can prevent CVD [73, 74]. The involvement of vitamin K in VC by the carboxylation of MGP has been confirmed in various animal studies that; MGP-knock out mice died within two months due to VC-induced rupturing of blood vessels followed by short stature, osteopenia, and fractures [75]. Sweatt et al. hypothesized that in rodents, a specific calcium-mediated and vitamin K-dependent Gla region in MGP protein is involved in binding bone morphogenetic protein-2 (BMP-2) that may link the age-related arterial calcification and low carboxylation of MGP [76]. Vitamin K antagonist warfarin has been shown to antagonize vitamin K-dependent carboxylation of MGP, leading to extensive VC [77]. However, vitamin K intake can suppress arterial calcification after treatment with

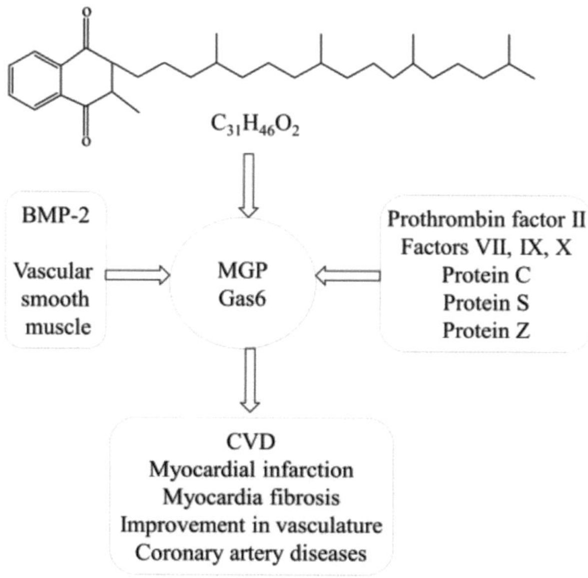

Figure 5.
A model showing the therapeutic potential of vitamin K against various cardiovascular diseases (CVD). MGP and Gas6 are vitamin K dependent proteins, which are activated by carboxylation. Subsequently, Gas6 exerts inhibitory effects on the apoptosis of endothelial cells and VSMCs, thus preventing angiosteosis and protecting blood vessels and improved various heart diseases.

warfarin in rats [78]. VKDPs, such as MGP and Gas-6, have the ability to protect the vasculature and have an essential role in blood coagulation by preventing tissue calcification and cell death in VSMCs and arterial vessel walls [79]. Several clinical observational studies have hypothesized that chronic dietary supplementation of both vitamins K1 and K2 may negatively correlate with risks of VC and CVD [79].

Additionally, subtypes of vitamin K2 (MK4 to MK9) have been examined in the Prospect–EPIC cohort, which consists of 16,057 women, aged 49–70 years old with no history of CVD [79]. This study concluded that a high intake of menaquinones (MK7, MK8, and MK9) could protect against CVD. However, this kind of protection has not been observed with vitamin K1 (phylloquinone) against CVD in other cohort observations [79, 80]. Notably, a Nurses' Health Study conducted with 72,874 female nurses aged between 38 and 65 years old has confirmed that vitamin K1 and lower risk of CVD is not significant because vitamin K1 intake may be a substitute marker for a healthy diet rather than an independent risk factor for CHD [81]. Nevertheless, data from National Health and Nutrition Examination Surveys examined the data of 5296 individuals with a minimum age of 50 years and concluded that vitamin K1 shows an independent assessment of high arterial pulse pressure [69]. In another prospective cohort study, 7216 participants were assessed by different types of vitamin K intake and mortality [82]. This study concluded that a high vitamin K intake is linked to the reduced risk of CVD in a Mediterranean population [82]. Vitamin K has shown promising results against vascular calcification in vitamin K-deficient individuals. Further research is justified to explore a relationship between vitamin K supplementation and the prevention of CVD.

7. Therapeutic role of vitamin K

CVD is a public health burden and a serious challenge to the health system throughout the world. CVD is a leading cause of death globally, with approximately

18 million deaths in 2015; the World Health Organization (WHO) forecasts that approximately 23.3 million deaths could occur from CVD by 2030 [83, 84]. The WHO has defined CVD as a "*group of illnesses that affect the heart and blood vessels*" [83]. These conditions include CHD and cerebrovascular disease. Scientific evidence has shown that factors related to nutrition have an important role in the development of CVDs and that these dietary factors may contribute to the differences in the morbidity and mortality from CVD seen in various regions of the world [83].

Different forms of vitamin K exist in the diet sourced from plants and animals [83, 85–88]. Vitamin K2 is usually a product of bacterial synthesis; however, meat, dairy, and fermented food products provide a minimal amount of vitamin K-2 [86, 88, 89]. Discovered in 1936, vitamin K has been known as an enzyme co-factor for the carboxylation of VKDPs [85, 86, 89]. Its key function in the synthesis of clotting factors in the liver has made the relationship between vitamin K and coagulation of blood a well-known phenomenon; however, recent studies are offering more insight into the diversity of functions associated with vitamin K [85, 86, 89, 90]. Many disease conditions related to the activities of vitamin K are now being described [89]. The carboxylation or activation of VKDPs requires vitamin K as a cofactor to the GGCX enzyme and occurs in the liver [83, 85, 86, 89]. The process converts specific Glu into calcium-binding Gla residues [86, 89, 90]. The uncarboxylated forms of the VKDPs are inactive, and carboxylation turns them into active and functioning proteins [86]. Some of the VKDPs that are carboxylated in the liver include clotting factors, such as factor II (prothrombin) and factor X [86]. Studies have confirmed that the process of activation of VKDPs also occurs outside the liver in smooth muscle cells. The extracellular matrix MGP protein is produced by smooth muscle cells and inhibits soft-tissue mineralization by binding to $ca2+$ ions to the vascular walls [44, 83, 85, 86]. MGP is a VKDP with Gla and serine residues. MGP is activated via carboxylation of the Gla residues, followed by phosphorylation of the serine residues [44, 86]. Vitamin K is essential to the carboxylation and phosphorylation of MGP as the enzyme co-factor [44, 86]. Carboxylation of MGP leads to its structural changes, which are very important for its ability to bind to calcium crystals [44]. Although MGP becomes inactivated when there is vitamin K deficiency and leads to VC, a high intake of vitamin K can reverse conditions [44, 83, 85, 86]. MGP secreted by chondrocytes and VSMCs has been shown to inhibit VC and was described as the most potent natural inhibitor of calcification in the human body [44]. Apart from inhibition of calcification, MGP has also been recognized as having the ability to reverse the calcification process [44]. The protection of MGP from VC occurs via its high binding affinity to new crystals of hydroxyapatite, which prevents their increase within the vascular wall [44]. MGP also stimulates arterial macrophages, leading to phagocytosis and apoptosis of the MGP-hydroxyapatite complex (**Figure 6**).

A sub-optimal level of vitamin K in the body is associated with an increased risk of adverse health outcomes, especially in adult and elderly populations. Studies have linked vitamin K deficiency with CVD, insulin resistance, and inflammation, as well as cognitive impairment [13, 83, 85, 87]. A lack of vitamin K has been shown to lead to an increased risk of calcification of blood vessels and CVD due to the presence of nonfunctioning Gla proteins [44, 83, 85, 86]. Vitamin K deficiency may cause increased calcium deposition in the walls of the blood vessels, leading to calcification of the coronary artery, and ultimately, CVD [13, 83]. Earlier observational studies have also established a relationship between low vitamin K intake with calcification of blood vessels; other observations have suggested that high vitamin K supplementation in the diet may reduce long-term CVD risks [13, 83, 85].

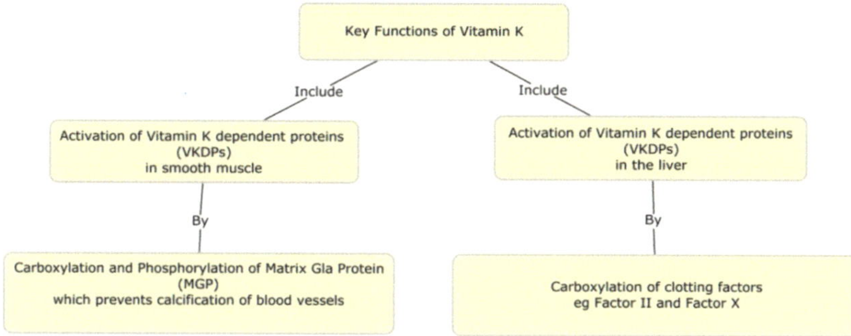

Figure 6.
Key functions of vitamin K.

An increased intake of dietary vitamin K is also associated with a decrease in the risk of all-cause mortality, as concluded by a study of a Mediterranean population with a high risk of CVD [83].

8. Vitamin K and inflammation

Inflammation is a recognized contributor to the progression and onset of diseases related to aging, such as osteoarthritis, CVD, and other similar diseases [85, 89]. The production of pro-inflammatory cytokines, such as tumor necrosis factor-alpha (TNF-α), C-reactive protein (CRP), and interleukin-6 (IL-6), has been found to be interfered by vitamin K. These findings were demonstrated in a cross-sectional study that showed a relationship among the high levels of vitamin K supplementation, the low levels of pro-inflammatory cytokines, and diminished inflammation in the body [85, 89]. Leptin hormone has a proinflammatory effect, menadione (VK3) has an apoptotic effect in Hepatocellular carcinoma through inhibiting leptin and through ROS generation which made VK3 a potential vitamin in preventing hepatocyte survival [91].

9. Conclusion

New insights about the activities of vitamin K and its crucial protective role in CVD development have emerged. Good association between dietary vitamin K2 and CVD is now clinically established. The role of inhibitory effect of statins in synthesis of vitamin K2 should be emphasized. However, there is still a need for prospective in-depth studies to improve the current understanding of this critical relationship and apply such knowledge to prevent CVD and improve its outcomes. Research should focus on understanding the function and regulation of new proteins that enhance or inhibit vascular calcification as well as the combination of vitamin D with other therapeutic drugs. Prospective studies may assess the vitamin K status using multiple biomarkers to provide insight on the relationship of vitamin K to vascular calcification and CVD.

Conflict of interest

All authors declare that there is no conflict of interest.

Author details

Meneerah A. Aljafary[1], Hussah Alshwyeh[1], Nada Alahmadi[1], Adeeb Shehzad[2], Huseyin Tombuloglu[3], Zagit Gaymalov[4], Abdelqader Homieda[1] and Ebtesam Al-Suhaimi[1*]

1 Biology Department, College of Sciences, Imam Abdulrahman Bin Faisal University, Dammam, Saudi Arabia

2 Department of Clinical Pharmacy Research, Institute for Research and Medical Consultations, Imam Abdulrahman Bin Faisal University, Dammam, Saudi Arabia

3 Genetics Research Department, Institute for Research and Medical Consultations, Imam Abdulrahman Bin Faisal University, Dammam, Saudi Arabia

4 Earlystage OÜ, Tallinn, Estonia

*Address all correspondence to: ealsuhaimi@iau.edu.sa

References

[1] Dam H. The antihaemorrhagic vitamin of the chick. The Biochemical Journal.1935; 29, 1273-1285.

[2] Vermeer C. Vitamin K: the effect on health beyond coagulation - an overview. Food Nutrition Research. 2012; 56.

[3] Danziger J. Vitamin K-dependent proteins, warfarin, and vascular calcification. Clinical journal of the American Society of Nephrology: CJASN.2008;3(5), 1504-1510.

[4] Schurgers, L. J., Uitto, J., Reutelingsperger, C. P.Vitamin K-dependent carboxylation of matrix Gla-protein: a crucial switch to control ectopic mineralization. Trends in Molecular Medicine. 2013;19, 217-226.

[5] Jaminon, A., Dai, L., Qureshi, A. R., Evenepoel, P., Ripsweden, J., Söderberg, M., Witasp, A., Olauson, H., Schurgers, L. J., &Stenvinkel, P. Matrix Gla protein is an independent predictor of both intimal and medial vascular calcification in chronic kidney disease. Scientific Reports.2020; 10, 6586.

[6] Laurance, S., Lemarié, C. A., Blostein, M. D. Growth arrest-specific gene 6 (gas6) and vascular hemostasis. Advances in nutrition (Bethesda, Md.).2012;3, 196-203.

[7] Xiao, H., Chen, J., Duan, L., & Li, S. Role of emerging vitamin K-dependent proteins: Growth arrest-specific protein 6, Gla-rich protein and periostin (Review). International journal of molecular medicine.2021; 47, 2.

[8] Booth, S. L., Al Rajabi, A. Determinants of vitamin K status in humans. Vitamins and hormones. 2008; 78, 1-22.

[9] Hartley, L., Clar, C., Ghannam, O., Flowers, N., Stranges, S., & Rees, K. Vitamin K for the primary prevention of cardiovascular disease. The Cochrane Database of Systematic Reviews.2015; 9, CD011148.

[10] van Ballegooijen, A. J., Beulens, J. W. The Role of Vitamin K Status in Cardiovascular Health: Evidence from Observational and Clinical Studies. Current Nutrition Reports. 2017; 6, 197-205.

[11] Halder, M., Petsophonsakul, P., Akbulut, A. C., Pavlic, A., Bohan, F., Anderson, E., Maresz, K., Kramann, R., &Schurgers, L. Vitamin K: Double Bonds beyond Coagulation Insights into Differences between Vitamin K1 and K2 in Health and Disease. International journal of molecular sciences.2019; 20, 896.

[12] Lees, J. S., Chapman, F. A., Witham, M. D., Jardine, A. G., & Mark, P. B. Vitamin K status, supplementation and vascular disease: a systematic review and meta-analysis. Heart (British Cardiac Society).2019;105, 938-945.

[13] Lee, S. E., Schulze, K, J., Cole, R. N., Wu, L. S. F., Yager, J. D., Groopman, J., Christian, P. and West, Jr., K. P. Biological systems of vitamin K: a plasma nutriproteomics study of subclinical vitamin k deficiency in 500 Nepalese children. OMICS Journal of Integrative Biology. 2015; 20, 214 –223.

[14] Booth SL, Broe KE, Gagnon DR, Tucker KL, Hannan MT, McLean RR, Dawson-Hughes B, Wilson PW, Cupples LA, Kiel DP. Vitamin K intake and bone mineral density in women and men. American Journal of Clinical Nutrition. 2012; 77, 512-516.

[15] Schurgers, L.J., Vermeer, C. Determination of phylloquinone and menaquinones in food; effect of food matrix on circulating vitamin K

concentrations. Haemostasis. 2000; 30, 298-307

[16] Koivu-Tikkanen, T.J., Ollilainen, V., Piironen, V.I. Determination of phylloquinone and menaquinones in animal products with fluorescence detection after postcolumn reduction with metallic zinc. J. Agric. Food Chem. 2000; 48, 6325-6331.

[17] Kamao M, Suhara Y, Tsugawa N, Uwano M, Yamaguchi N, Uenishi K, Ishida H, Sasaki S, Okano T. Vitamin K content of foods and dietary vitamin K intake in Japanese young women. J Nutr Sci Vitaminol (Tokyo). 2007, 53(6):464-470.

[18] Elder SJ, Haytowitz DB, Howe J, Peterson JW, Booth SL. Vitamin k contents of meat, dairy, and fast food in the U.S. Diet. Journal of Agriculture Food Chemistry. 2006; 54, 463-467.

[19] Homeida A. M., Hussein F.M., AL-jumaah R.S., Alhaidary M. A., Alshaikh M.A., Al-bokhadaim I, Gar ElnabiA. Serum vitamin k1, a and e levels in dromedary camels (camelus dromedaries) with notes on vitamin k1 in other animals. Journal of Camel Practice and Research. 2010; 17, 14-18.

[20] Thamratnopkoon S, Susantitaphong P, Tumkosit M, Katavetin P, Tiranathanagul K, Praditpornsilpa K, Eiam-Ong S. Correlations of plasma desphosphorylateduncarboxylated Matrix Gla protein with vascular calcification and vascular Stiffness in chronic kidney disease. Nephron. 2017; 135,167-172.

[21] Geleijnse JM, Vermeer C, Grobbee DE, Schurgers LJ, Knapen MH, van der Meer IM, Hofman A, Witteman JC. Dietary Intake of Menaquinone is Associated with a Reduced Risk of Coronary Heart Disease: The Rotterdam Study. Journal of Nutrition. 2004; 134, 3100-3105.

[22] Haugsgjerd TR, Egeland GM, Nygård OK, Vinknes KJ, Sulo G, Lysne V, Igland J, Tell GS. Association of dietary vitamin K and risk of coronary heart disease in middle-age adults: the Hordaland Health Study Cohort. BMJ Open. 2020; 10,e035953.

[23] Gast, N M, Sluijs, I, Bots, M I, Beulens, J M J Geleijnse, J M Witteman, J C, Grobbee, D E, Peeters, PHM van der Schouw YT. A high menaquinone intake reduces the incidence of coronary heart disease. Nutrition Metabolism Cardiovascular Diseases. 2009; 19, 504-510.

[24] Knapen MHJ, Braam LAKJLM, Drummen NE, Bekers O, Hoeks APG, Vermeer C. Menaquinone-7 supplementation improves arterial stiffness in healthy postmenopausal women. A double-blind randomised clinical trial. Thrombosis Haemostasis. 2015; 113,1135-44

[25] Fathima, H. M., Ganapathy, D., Subhashree, R., Rakshagan, V. Knowledge and Awareness of Vitamin K Therapy among Dental Students. Journal of Pharmaceutical Research International, 2020; 38-47.

[26] Gordeladze, J. O. Vitamin K2: A Vitamin that Works like a Hormone, Impinging on Gene Expression. In Cell Signalling-Thermodynamics and Molecular Control. Intech Open. 2019.

[27] Lanham, S. A., Cagampang, F. R., Oreffo, ROC. Maternal High-Fat Diet and Offspring Expression Levels of Vitamin K–Dependent Proteins. Endocrinology, 2014; 155(12), 4749-4761.

[28] Gundberg, C. M., Lian, J. B., Booth, S. L. Vitamin K-dependent carboxylation of osteocalcin: friend or foe? Advances in Nutrition. 2012; 3, 149-157.

[29] O'Connor, E. M., Durack, E. Osteocalcin: The extra-skeletal role of a

vitamin K-dependent protein in glucose metabolism. Journal of Nutrition & Intermediary Metabolism. 2017; 7, 8-13.

[30] Al-Suhaimi, E. A., Al-Jafary, M.A. Endocrine roles of vitamin K-dependent-osteocalcin in the relation between bone metabolism and metabolic disorders. Reviews in Endocrine and Metabolic Disorders. 2020; 21, 117-125.

[31] Sato, T., Schurgers, L. J., Uenishi, K. Comparison of menaquinone-4 and menaquinone-7 bioavailability in healthy women. Nutrition Journal. 2012; 11(1), 93.

[32] Scheiber, D., Veulemans, V., Horn, P., Chatrou, M.L., Potthoff, S. A., Kelm, M., Westenfeld, R. (2015). High-dose menaquinone-7 supplementation reduces cardiovascular calcification in a murine model of extraosseous calcification. Nutrients. 2015; 7, 6991-7011.

[33] Rishavy MA, Berkner KL. Vitamin K oxygenation, glutamate carboxylation, and processivity: defining the three critical facets of catalysis by the vitamin K–dependent carboxylase. Advances in nutrition. 2012; 3, 135-48.

[34] Fusaro M, Gallieni M, Porta C, Nickolas TL, Khairallah P. Vitamin K effects in human health: new insights beyond bone and cardiovascular health. Journal of Nephrology. 2020;33, 239-49.

[35] Tie JK, Stafford DW. Structural and functional insights into enzymes of the vitamin K cycle. Journal of Thrombosis and Haemostasis. 2016;14, 236-47.

[36] Tarento TD, McClure DD, Talbot AM, Regtop HL, Biffin JR, Valtchev P, Dehghani F, Kavanagh JM. A potential biotechnological process for the sustainable production of vitamin K1. Critical Reviews in Biotechnology. 2019;39, 1-9.

[37] Roumeliotis S, Dounousi E, Eleftheriadis T, Liakopoulos V. Association of the inactive circulating matrix Gla protein with vitamin K intake, calcification, mortality, and cardiovascular disease: a review. International Journal of Molecular Sciences. 2019;20, 628.

[38] Villa-Bellosta R, Millan A, Sorribas V. Role of calcium-phosphate deposition in vascular smooth muscle cell calcification. American Journal of Physiology-Cell Physiology. 2011;300,210-20.

[39] Kalampogias A, Siasos G, Oikonomou E, Tsalamandris S, Mourouzis K, Tsigkou V, Vavuranakis M, Zografos T, Deftereos S, Stefanadis C, Tousoulis D. Basic mechanisms in atherosclerosis: the role of calcium. Medicinal Chemistry. 2016;12, 103-13.

[40] Poterucha TJ, Goldhaber SZ. Warfarin and vascular calcification. The American journal of medicine. 2016;129, 635-e1.

[41] Gallieni M, Fusaro M. Vitamin K and cardiovascular calcification in CKD: is patient supplementation on the horizon. Kidney international. 2014;86, 232-4.

[42] Speer MY, Yang HY, Brabb T, Leaf E, Look A, Lin WL, Frutkin A, Dichek D, Giachelli CM. Smooth muscle cells give rise to osteochondrogenic precursors and chondrocytes in calcifying arteries. Circulation research. 2009;104, 733-41.

[43] Shea MK, O'Donnell CJ, Hoffmann U, Dallal GE, Dawson-Hughes B, Ordovas JM, Price PA, Williamson MK, Booth SL. Vitamin K supplementation and progression of coronary artery calcium in older men and women. The American journal of clinical nutrition. 2009;89, 1799-807.

[44] Roumeliotis, S., Dounousi, E., Eleftheriadis, T. and Liakopoulos, V.: Association of the inactive circulating matrix gla protein with vitamin k intake, calcification, mortality, and cardiovascular disease: a review. International Journal of Molecular Sciences. 2019; 20, 1- 27.

[45] Gaw A, Packard MRCPath C, James Shepherd, MB, ChB, PhD, FRCPath, FRCP (Glasg). Statins: The HMG CoA reductase inhibitors in perspective. CRC Press; 1999. 224.

[46] Gaw A. Statins in General Practice: Pocketbook. CRC Press; 2001. 112 p.

[47] Estren MJ. Statins: Miracle or Mistake? Ronin Publishing; 2014. 200 p.

[48] Torjesen I. Statins are overprescribed for primary prevention, study suggests. BMJ. 2018. p. k5110.

[49] Ilana B. Richman, Joseph S. Ross. Weighing the Harms and Benefits of Using Statins for Primary Prevention: Raising the Risk Threshold. Ann Intern Med. 2019;170, 62-63.

[50] Demasi M. Statin wars: have we been misled about the evidence? A narrative review. British Journal of Sports Medicine. 2018; 52, 905-9.

[51] Yebyo HG, Aschmann HE, Puhan MA. Finding the Balance Between Benefits and Harms When Using Statins for Primary Prevention of Cardiovascular Disease: A Modeling Study. Annals of Intern Med. 2019;170, 1-10.

[52] U. S. Food and Drug Administration, Office of the Commissioner. Controlling Cholesterol with Statins [Internet]. 2020 [cited 2021 Jan 27]. Available from: https://www.fda.gov/consumers/consumer-updates/controlling-cholesterol-statins

[53] Ravnskov U, Okuyama H, Sultan S. Serious bias in 20 year follow-up study of statin trial. BMJ. 2017; 359, 4906.

[54] Okuyama H, Langsjoen PH, Ohara N, Hashimoto Y, Hamazaki T, Yoshida S, et al. Medicines and Vegetable Oils as Hidden Causes of Cardiovascular Disease and Diabetes. Pharmacology. 2016; 98, 134-70.

[55] Okuyama H, Langsjoen PH, Hamazaki T, Ogushi Y, Hama R, Kobayashi T, et al. Statins stimulate atherosclerosis and heart failure: pharmacological mechanisms. Expert Rev Clinical Pharmacology. 2015;8,189-99.

[56] Hashimoto Y, Okuyama H. Statins Cause Lifestyle-Related Diseases-Biochemical Mechanism. Endocrinology Diabetes Research. 2017; 10:2.

[57] Yandrapalli S, Malik A, Guber K, Rochlani Y, Pemmasani G, Jasti M, et al. Statins and the potential for higher diabetes mellitus risk. Expert Review in Clinical Pharmacology. 2019;12, 825-830.

[58] Culver AL, Ockene IS, Balasubramanian R, Olendzki BC, Sepavich DM, Wactawski-Wende J, et al. Statin use and risk of diabetes mellitus in postmenopausal women in the Women's Health Initiative. Archives of International Medicines. 2012;172, 144-52.

[59] Geleijnse JM, Vermeer C, Grobbee DE, Schurgers LJ, Knapen MHJ, van der Meer IM, et al. Dietary intake of menaquinone is associated with a reduced risk of coronary heart disease: the Rotterdam Study. Journal of Nutrition. 2004;134, 3100-5.

[60] Cederberg H, Stančáková A, Yaluri N, Modi S, Kuusisto J, Laakso M. Increased risk of diabetes with statin treatment is associated with impaired insulin sensitivity and insulin secretion: a 6year follow-up study of the METSIM cohort. Diabetologia. 2015; 58,1109-17.

[61] Gordeladze J. Vitamin K2: Vital for Health and Wellbeing. BoD – Books on Demand; 2017. 338 p.

[62] Booth SL, Tucker KL, Chen H, Hannan MT, Gagnon DR, Cupples LA, et al. Dietary vitamin K intakes are associated with hip fracture but not with bone mineral density in elderly men and women. American Journal of Clinical Nutrition. 2000; 71, 1201-8.

[63] Shiraki M, Shiraki Y, Aoki C, Miura M. Vitamin K2 (menatetrenone) effectively prevents fractures and sustains lumbar bone mineral density in osteoporosis. Journal of Bone Mineral Research. 2000;15, 515-21.

[64] Villa JKD, Diaz MAN, Pizziolo VR, Martino HSD. Effect of vitamin K in bone metabolism and vascular calcification: A review of mechanisms of action and evidences. Critical Review in Food Science Nutrition. 2017; 57, 3959-3970.

[65] van Ballegooijen, A. J., Beulens, J. W. The Role of Vitamin K Status in Cardiovascular Health: Evidence from Observational and Clinical Studies. Current Nutrition Reports. 2017; 6, 197-205.

[66] Schurgers LJ, Cranenburg EC, Vermeer C. Matrix Gla-protein: the calcification inhibitor in need of vitamin K. Thrombosis Haemostasis. 2008;100, 593-603.

[67] Marles RJ, Roe AL, Oketch-Rabah HA. US Pharmacopeial Convention safety evaluation of menaquinone-7, a form of vitamin K. Nutrition Reviews. 2017; 75, 553-578.

[68] Food and Nutrition Board, Institute of Medicine, Dietary Reference Intakes for vitamin A, vitamin K, Arsenic, Boron, Chromium, Copper, Iodine, Iron, Manganese, Molybdenum, Nickel, Silicon, Vanadium, and Zinc, The National Academic Press, Washington, DC, USA, 2001.

[69] Vlasschaert C, Goss CJ, Pilkey NG, McKeown S, Holden RM. Vitamin K Supplementation for the Prevention of Cardiovascular Disease: Where Is the Evidence? A Systematic Review of Controlled Trials. Nutrients. 2020;12, 2909.

[70] Heart Disease and Stroke Statistics—2020 Update: A Report From the American Heart Association. Circulation, 2020, 141, 9.

[71] Scarborough P, Morgan RD, Webster P, Rayner M. Differences in coronary heart disease, stroke and cancer mortality rates between England, Wales, Scotland and Northern Ireland: the role of diet and nutrition. BMJ Open. 2011;1, e000263.

[72] DiNicolantonio JJ, Bhutani J, O'Keefe JH. The health benefits of vitamin K. Open Heart. 2015; 2, e000300.

[73] Gast GC, de Roos NM, Sluijs I, Bots ML, Beulens JW, Geleijnse JM, Witteman JC, Grobbee DE, Peeters PH, van der Schouw YT. A high menaquinone intake reduces the incidence of coronary heart disease. Nutrition Metabolism & Cardiovascular Diseases. 2009;19, 504-10.

[74] Shea MK, O'Donnell CJ, Hoffmann U, Dallal GE, Dawson-Hughes B, Ordovas JM, Price PA, Williamson MK, Booth SL. Vitamin K supplementation and progression of coronary artery calcium in older men and women. American Journal of Clinical Nutrition. 2009; 89, 1799-807.

[75] Luo G, Ducy P, McKee MD, Pinero GJ, Loyer E, Behringer RR, et al. Spontaneous calcification of arteries and cartilage in mice lacking matrix GLA protein. Nature. 1997; 386, 78-81.

[76] Sweatt A, Sane DC, Hutson SM, Wallin R. Matrix Gla protein (MGP) and bone morphogenetic protein-2 in aortic calcified lesions of aging rats.

Journal of Thrombosis and Haemostasis. 2003; 1, 178-85.

[77] Price PA, Faus SA, Williamson MK. Warfarin causes rapid calcification of the elastic lamellae in rat arteries and heart valves. Arteriosclerosis, Thrombosis, and Vascular Biology. 1998;18,1400-7.

[78] Schurgers LJ, Spronk HM, Soute BA, Schiffers PM, DeMey JG, Vermeer C. Regression of warfarin-induced medial elastocalcinosis by high intake of vitamin K in rats. Blood. 2007;109, 2823-31.

[79] Gast GC, de Roos NM, Sluijs I, Bots ML, Beulens JW, Geleijnse JM, et al. A high menaquinone intake reduces the incidence of coronary heart disease. Nutrition, Metabolism, and Cardiovascular Diseases: NMCD. 2009;19, 504-10.

[80] Erkkila AT, Booth SL, Hu FB, Jacques PF, Lichtenstein AH. Phylloquinone intake and risk of cardiovascular diseases in men. Nutrition, Metabolism, and Cardiovascular Diseases: NMCD. 2007;17, 58-62.

[81] Erkkila AT, Booth SL, Hu FB, Jacques PF, Manson JE, Rexrode KM, et al. Phylloquinone intake as a marker for coronary heart disease but not stroke in women. European Journal of Clinical Nutrition. 2005; 59,196-204.

[82] Juanola-Falgarona M, Salas-Salvado J, Martinez-Gonzalez M, Corella D, Estruch R, Ros E, et al. Dietary intake of vitamin K is inversely associated with mortality risk. Journal of Nutrition. 2014;144, 743-50.

[83] Hartley, L., Clar, C., Ghannam, O., Flowers, N., Stranges, S. Rees, K. Vitamin K for the primary prevention of cardiovascular disease. Cochrane Database of Systematic Reviews. 2015; CD011148.

[84] Wen, L., Chen, J., Duan, L. and Li, S.: Vitamin K dependent proteins involved in bone and cardiovascular health (Review). Molecular Medicine Reports. 2018; 18, 3-15.

[85] Harshman, S.G. Shea, M.K. The role of vitamin K in chronic aging diseases: inflammation, cardiovascular disease and osteoarthritis. Current Nutrition Reports. 2016; 5, 90-98.

[86] Ballegooijen, A.J. Beulens, J. W. The role of vitamin K status incardiovascular health: evidence from observational and clinical studies. Current Nutrition Reports. 2017; 6:197-205.

[87] Sim et al.: The effects of vitamin k-rich green leafy vegetables on bone metabolism: a 4- week randomized controlled trial in middle-aged and older individuals. Bone Reports. 2020; 12: 1-7.

[88] Walther, B., Karl, J. P., Booth, S. L. and Boyaval, P. Menaquinones, bacteria, and the food supply: the relevance of dairy and fermented food products to vitamin K requirements. Advances in Nutrition. 2013; 463-473.

[89] Simes, D.C., Viegas, C.S.B., Araújo, N. Marreiros, C. Vitamin K as a diet supplement with impact in human health: current evidence in age-related diseases. Nutrients. 2020; 138.

[90] Sankar, M. J., Kumar, A. C. P., Agarwal, A. T. R. and Paul, V. K.: Vitamin K prophylaxis for prevention of vitamin K deficiency bleeding: a systematic review. Journal of Perinatology. 2016; 36, 29-34.

[91] Al-Suhaimi E. Molecular mechanisms of leptin and pro-apoptotic signals induced by menadione in HepG2 cells. Saudi Journal of Biological Sciences. 2014; 21, 6, 582-588. doi:10.1016/j.sjbs.2014.03.002

Chapter 4

Menaquinone-7: Wide Ranging Physiological Relevance in Muscle and Nerve Health

Dilip Mehta, Anselm de Souza and Shashank S. Jadhav

Abstract

Menaquinone-7 plays a significant role in cardiovascular and bone health. In recent times there is a growing interest in understanding the role of Menaquinone-7 in health and diseases. Several population-based studies have reported specific health effects of the long-chain menaquinones, notably MK-7, MK-8, and MK-9. There are several epidemiological studies, clinical trials, along with *in vivo* and *in vitro* studies confirming the role of Menaquinone-7 in health and diseases. More recently, research group at Synergia Life Sciences has discovered a wider role for Menaquinone-7 in energy homeostasis (VO_{2max}), peripheral neuropathy, muscle cramps and mitochondrial respiration not only through improvement of the electron transport but also the perfusion improving oxygen availability. In the current chapter, the authors have discussed the wider physiological role of Menaquinone-7 highlighting the recent research with Menaquinone-7 in the areas of Muscle and Nerve Health.

Keywords: Menaquinone-7, muscle cramps, peripheral neuropathy, bone health, cardiovascular, insulin resistance, deficiency, catabolism, SXR, energy homeostasis

1. Introduction

Menaquinone-7 belongs to Vitamin K group. The two general categories of vitamin K are Phylloquinone (Vitamin K1) and Menaquinones, also referred as MK-n (clinical nomenclature Vitamin K2-n) having side chains with 4–12 prenyl units (MK-n where n stands for the number of isoprenoid units, MK-4 to MK-12) 9 (**Figure 1**). Their physiological and patho- physiological roles are specific.

2. Biological activity of Menaquinone-7

Vitamin K was discovered by the Danish scientist, Henrik Dam, in the 1930s. Dam's discovery was during his quest to understand chicken's cholesterol metabolism by feeding them a diet free of sterols and low in fat [1]. This reduced their intake of fat-soluble vitamin K, resulting in chickens developing large subcutaneous and intramuscular hemorrhages. This initial finding led to isolating, identifying, and characterizing the structure of vitamin K and its importance as an anti-haemorrhagic agent. Of the many metabolic processes related to vitamin K

(I)

(II)

(III)

(IV)

(V)

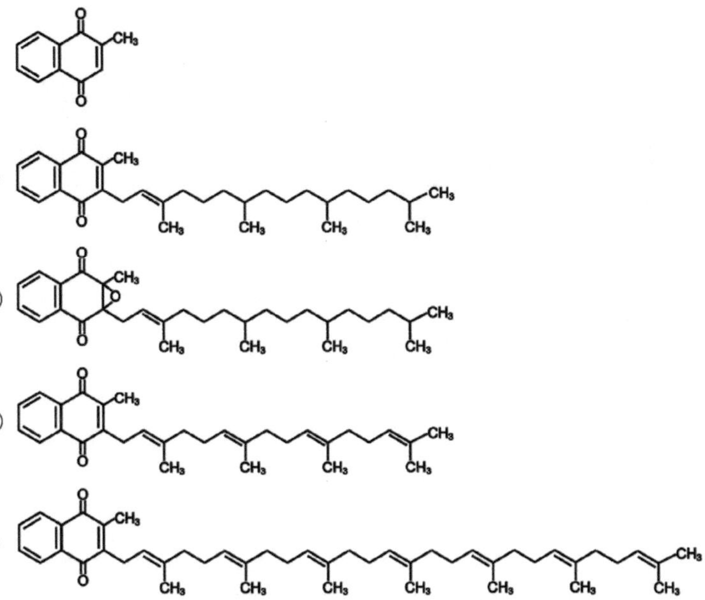

Figure 1.
Chemical structures of various isoforms of Vitamin K. chemical structures of some K vitamins and metabolites. Nomenclature: Chemical name and IUPAC name and abbreviation in brackets: (I) 2-methyl1,4-naphthoquinone (Menadione; K3), (II) 2-methyl3-phytyl-1,4-naphthoquinone (Phylloquinone; K1), (III) 2-methyl-3-phytyl-1,4-naphthoquinone-2,3-epoxide (Phylloquinone epoxide; K1O), (IV) 2-methyl-3-geranyl-geranyl1,4-naphthoquinone (Menaquinone-4; MK-4), (V) 2-methyl-3-farnesylgeranylgeranyl-1,4-naphthoquinone (Menaquinone-7; MK-7).

deficiency, bleeding remains the potentially most serious generally known consequence. However, the role of vitamin K's impact on osteoporosis and its inhibitory role in arterial calcification and vascular biology is now recognized in general populations. It is axiomatic that these metabolisms require vitamin K for γ-carboxylation and that this step is essential to their proper functioning. However, there are many other functions of vitamin K recently discovered that seem to be independent of its classical co-factor function. Vitamin K's metabolic effects, e.g., ameliorating effect on peripheral neuropathy, cramps, autonomic nervous system, improving perfusion, etc., remain unexplained. Additionally, vitamin K also acts as a ligand for the receptor SXR, the steroid and xenobiotic sensing nuclear receptor (SXR), which is a transcriptional regulator of the cytochrome P450 gene CYP3A4.

Over the years, the understanding of the vitamin K family has evolved, with the recognition of two primary forms of vitamin K- vitamin K1 (phylloquinone) and vitamin K2 (menaquinones). All K-vitamins have same function, but they exhibit differences in bioavailability and bioactivity. Vitamin K2, the main storage form in animals, has several subtypes, which differ in isoprenoid sidechain length. These vitamin K2 homologs are called menaquinones and are characterized by the number of isoprenoid residues in their side chains. Menaquinones are abbreviated as MK-n, where M stands for menaquinone, the K stands for vitamin K, and the n signifies the number of isoprenoid side chain residues. MK-4 and MK-7 are the two prominent menaquinones in human nutrition. MK-7 and other long-chain menaquinones are different from MK-4 in that they are not produced by human tissue but are generated by bacteria in the gut. The available information suggests that a range of vitamin K2 analogues are present as a mixture in several foods, e.g., in sauerkraut, hard cheese, soft cheese and curd cheese [2]. These foods have a long history of consumption by humans as basic foods.

Diet (Natto and cheeses) rich in menaquinones (primarily MK-7) is safe and requires no systemic validation for toxicity. However, because of the role of vitamin K (K1 and K2) in blood coagulation and potential health benefits, there has been considerable effort to elucidate the mechanism of action of menaquinones, primarily MK-7. There is no known toxicity associated with high doses (dietary or supplemental) of the phylloquinone (vitamin K1) or menaquinones (vitamin K2) forms of vitamin K. In several human studies, Natto food, known to contain MK-7, has been investigated for its health benefits.

3. Menaquinone-7 safety

The adverse effects of menaquinones, including Menaquinone-7 has been investigated in several animal and *in vitro* toxicity studies. Findings from animal studies for acute, chronic, and genotoxicity and *in vitro* studies for mutagenicity and carcinogenicity showed no significant risks associated with exposure to menaquinones [3]. The absence of adverse effects or death suggest that the minimum lethal dose of Menaquinone-7 is greater than 2000 mg/kg bw [4].

The European Union has permitted the use of Menaquinone-7 as a source of vitamin K for nutritional purposes in foodstuffs. The European Food Safety Authority (EFSA, 2008) [5] examined the safety of Menaquinone-7. The chemistry, nomenclature, dietary sources, intake levels, and pharmacokinetics of menaquinones, and data of nonclinical toxicity and on clinical outcomes related to safety (adverse events) was extensively reviewed by US Pharmacopeia Convention [6] and by the Institute of Medicine (IOM, 2000). The report considers menaquinone as an active form of vitamin K [7].

4. Menaquinone-7 in diet

Schurgers et al. [2] have studied levels of Menaquinone-7 in many food products globally and found that the Menaquinone-7 levels are quite negligible in all the food products except Natto, a staple food in Eastern Japan which contains almost 998 mcg of Menaquinone-7 per 100 gm of Natto. The investigators also found small amounts of Menaquinone-7 in natural cheese. Researchers at Synergia Life Sciences have undertaken a study where a number of Indian food products were examined for the levels of Menaquinone-7. The foods tested had particularly included fermented foods consumed by Indians at large. It was observed that the regularly consumed food including fermented foods lack in Menaquinone-7. So, it can be said that Menaquinone-7 is negligible in Indian diet.

The only rich source of Menaquinone-7 is Natto which contains early 900 mcg of Menaquinone-7 in 100 gm's breakfast [2, 8] and different types of cheese [9] though in small amounts. The various common 18 varieties of Dutch cheeses and 13 varieties of European cheeses contain approximately on an average 1.14 and 1.36 mcg Menaquinone-7 per 100 gm of cheese respectively [9]. The hard cheese, soft cheese and curd cheese from Netherlands contains approximately 1.3, 0.5 and 0.3 mcg Menaquinone-7 per 100 gm cheese respectively [2]. Processed cheese from Japan reported 0.3 mcg Menaquinone-7 per 100 gm cheese [10].

Serum concentrations of Menaquinone-7 are higher in frequent natto eaters. Natto is a popular breakfast item used more widely in win Eastern Japan, as compared to Western Japan. The study by Kaneki *et al*. reports an average serum Menaquinone-7 concentration of 5.26 ng/ml in Eastern Japanese women (Tokyo), 1.22 ng/ml in the Western Japanese women (Hiroshima) and 0.37 ng/ml in British women (London) [11]. The serum concentrations in British women are negligible since they do not consume Natto, but their diet may include cheese which contributes to the small amounts of Menaquinone-7 in their serum.

Globally speaking Menaquinone-7 is negligible in diet. It is true that many bacteria that populate microbiota of the human intestine synthesize Menaquinones. However, it is realized that in the small intestine bacterial growth availing Menaquinone-7 is limited by the rapid transit times. Most synthesis of Menaquinones occur in the large intestine. Shearer *et al.* [12] and Suttie *et al.* [13] have examined the evidence of the contribution of gut menaquinones and concluded that while they do contribute to the human nutrition but not significantly. Karl JP *et al.* have shown that total Menaquinone (Menaquinone-4 to Menaquinone-13) concentration in human gut is highly variable. They measured total daily excretion of menaquinones in feces. The median total daily excretion of menaquinones in feces was 850 nmol/d but was highly variable (Range: 64–5358 nmol/day) [14].

5. Role of Menaquinone-7 in various diseases

5.1 Cardiovascular diseases

Geleijnse et al. [15] studied 4807 men and women of aged 55 yrs. for 10 years to assess the association of dietary intake of K1 and K2 with aortic calcification, CVD, and total mortality. They concluded that "When consuming daily 45 mcg dietary K2, you have: 50% reduction of arterial calcification, 50% reduction of cardiovascular death, 25 % reduction of all-cause mortality as compared to low intake of dietary K2!"

Gast et al. [16] studied 16,057 women, aged 49–70 years and free of cardiovascular diseases (at baseline) for 8.1 ± 1.6 years. The intake of vitamin K1 was 211.7 ± 100.3 mcg/d and of vitamin K2 intake was 29.1 ± 12.8 mcg/d. They concluded that there is inverse association of vitamin K2 with CHD with reduction of 9.1% per 10 mcg/day. They also found out that vitamin K1 is not related to CHD.

The publication of the above two epidemiological studies, viz. Geleijnse *et al.* and Gast *et al.* Study, has expanded interest the investigations of various beneficial health effects of Menaquinone-7.

5.2 Bone health

Knapen *et al.* [17] investigated the effects of low-dose Menaquinone-7 on bone health in healthy postmenopausal women. Menaquinone-7 intake significantly improved vitamin K status and decreased the age-related decline in Bone Mineral Content (BMC) and Bone Mineral Density (BMD) at the lumbar spine and femoral neck. In another placebo-controlled study, the authors investigated the effect of Menaquinone-7 on BMD and found out that Menaquinone-7 preserves trabecular bone structure at the tibia along with decrease in undercarboxylated osteocalcin (ucOC) [18]. In another clinical study Kanellakis *et al.* [19] assessed the effect of dairy products enriched with calcium, vitamin D3, and Menaquinone-7 on parameters of bone metabolism in postmenopausal women following a 12-month intervention. The study revealed more favorable changes in bone metabolism and bone mass indices for the Vitamin K2 supplemented groups. Van Summeran *et al.* [20] studied the effect of 45 mcg Menaquinone-7 on the circulating levels of undercarboxylated osteocalcin (ucOC) and carboxylated osteocalcin (cOC) along with Menaquinone-7 levels in healthy prepubertal children. They showed that the levels of Menaquinone-7 increased with the supplementation of Menaquinone-7 as compared to baseline levels.

Spronk HMH *et al*. in 2003 conducted an *in vivo* study for assessing tissue specific utilization of Vitamin K2 which resulted in prevention of arterial calcification in warfarin-treated rats. It was shown that the utilization of Vitamin K2 was more efficient in the aorta as compared to other tissues [21].

In a study by Yamaguchi *et al*., the authors have shown the anabolic effect of Menaquinone-7 on bone tissue and osteoblastic MC3T3-E1 cells *in vitro* [22]. Min Zhu *et al*. have shown that Menaquinone-7 has a stimulatory effect on bone tissue and osteoblastic SAOS-2 cells *in vitro* [23]. These studies suggest the role of Menaquinone-7 in osteoblastic bone formation. Recently, it has also been shown that Menaquinone-7 protects osteoblasts from oxidative stress and has beneficial effects on proliferation, differentiation, and mineralization of osteoblasts [24].

5.3 Insulin resistance

A decade-long study of 38,094 Dutch males and females aged 20–70 found that the quartile of participants who consumed the most dietary Vitamin K were 20% less likely to develop type 2 diabetes than the quartile with the lowest intake of Vitamin K [25]. Vitamin K2 is linked to lowered risk of developing type 2 diabetes and also a stronger relationship exists for Vitamin K2 intake. The risk of developing type 2 diabetes drops for every 10 mcg (0.01 mg) increase in Vitamin K2 intake. In this study participants with the highest intake of K2 consumed 250–360 mcg (0.25–0.36 mg)/day. Thus, higher intake of Vitamin K2 is linked to lower diabetes risk.

In an attempt to better understand how Vitamin K2 improves insulin sensitivity, researchers from S. Korea studied 42 healthy male volunteers. Participants were either given 30 mg (30,000 mcg) of Vitamin K2 or a placebo each day for 4 weeks. Vitamin K2 supplementation significantly increased insulin sensitivity and seemed to be related to increased carboxylation (activation) of osteocalcin. Researchers concluded that Vitamin K2 can help regulate glucose metabolism by activating osteocalcin, an endocrine hormone that increases insulin sensitivity in humans [26].

Research has shown that for elderly men Vitamin K slows the development of insulin resistance [27]. The researchers concluded that Vitamin K2 plays a potentially beneficial role in reducing the progression of insulin resistance amongst elderly men.

6. Recent research

6.1 Energy homeostasis (VO_{2max})

In a recent randomized controlled trial, McFarlin et al. investigated the effects of dietary supplementation of Menaquinone-7 on cardiovascular responses to a graded cycle ergometer test. Menaquinone-7 supplementation was associated with a 12% increase in maximal cardiac output, with a trend toward an increase in heart-rate AUC. No significant changes occurred in stroke volume [28]. **Figure 2** demonstrates that Menaquinone-7 treatment was associated with increased cardiac output, stroke volume, heart rate, and decreased blood lactate. Overall, these changes are consistent with increase maximal cardiovascular performance with oral Menaquinone-7 supplementation.

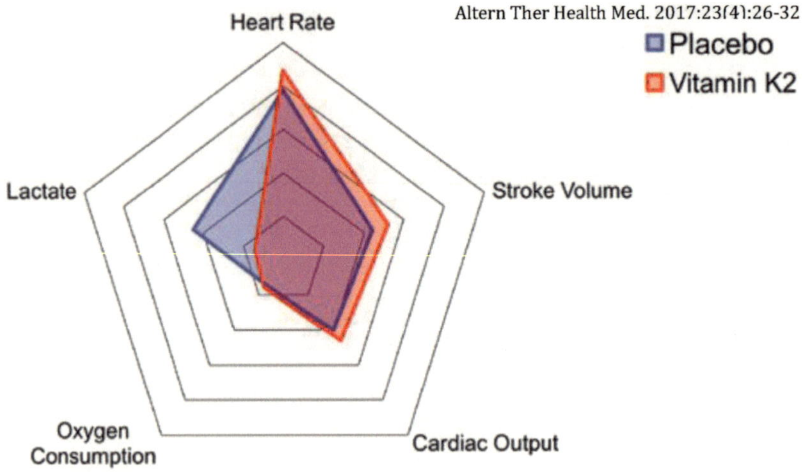

Figure 2.
To visualize the 5 outcome variables (heart rate, stroke volume, cardiac output, oxygen consumption, and blood lactate) on the same scale all data maximal response data after 8 weeks of treatment with either a vitamin K2 (red) or control (rice flour; blue) were normalized using a Log10 adjustment. Plotted values represent increments of a Log10 scale consuming a specific supplement.

6.2 Mitochondrial respiration

Synergia research group has identified Menaquinone-7's pivotal role in mitochondrial ATP generation by acting as a mitochondrial electron transport carrier, thus participating in the energy cycle of the cell. In human cell experiments, it has been shown that the cells' maximum capacity to generate energy, defined as the reserve energy, increases by 30–40% with Menaquinone-7, thus, identifying the role of Menaquinone-7 in redox cycle by transporting electrons in electron transport chain and also mitochondrial generation of ATP (**Figure 3**). This dual role of Menaquinone-7 is especially important to the aging geriatric population and athletes in their need of a greater oxygen supply for the oxidative phosphorylation.

In another *in vitro* study, Menaquinone-7 rescued mitochondrial defects in numerous conditions that affect mitochondrial function. Menaquinone-7 was also effective at improving systemic locomotion defects in fully developed adult pink1 and parkin mutant flies. Menaquinone-7 did not affect mitochondrial remodeling

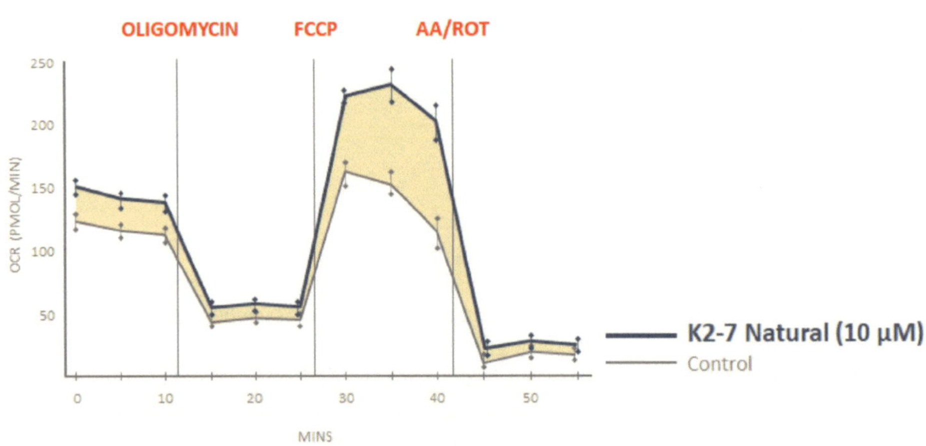

Figure 3.
Mitochondrial respiration: Test sequence in sea horse XF-96 platform.

directly, but by increasing Electron Transfer Chain efficiency, it contributed to the proton motif force that facilitates ATP production. Menaquinone-7 may thus constitute a promising compound to treat mitochondrial pathology, also in Parkinson's disease (PD) patients suffering from Pink1 or Parkin deficiency [29]. A clinical study has been proposed to investigate the potential effects of Menaquinone-7 in genetically determined PD with mitochondrial dysfunction [30].

6.3 Anti-inflammatory

Chronic inflammation is considered an underlying pathology of many diseases that remain poorly understood and treated. Several important chronic diseases with an inflammatory background have been associated with vitamin K deficiency. These include cystic fibrosis, inflammatory bowel disease, pancreatitis, chronic kidney disease and osteoporosis [31, 32]. Circulatory markers of low-grade inflammation such as tumor necrosis factor-alpha (TNF-α), interleukin-1 alpha (IL-1α), and interleukin-1 beta (IL-1β) positively correlate with endothelial damage, atheroma formation, cardiovascular disease, and aging. Menaquinone-7 can modulate immune and inflammatory reactions in the dose–response inhibition of TNF-α, IL-1α, and IL-1β gene expression and protein production [33]. These findings highlight the anti-inflammatory properties of Menaquinone-7, elucidating the anti-inflammatory mechanism of Menaquinone-7 and in establishing the potential biomarker targets in clinical testing of the role of Menaquinone-7 in cardiovascular health as well as other chronic degenerative conditions.

6.4 Muscle health

Vitamin K deficiency impacts neuromuscular and vascular function, thus affecting the physical functioning. Vitamin K has a function in promoting vascular smooth muscle differentiation [34]. As disabilities in patients are directly related to muscle strength and physical performance, therefore it is crucial to focus on muscle strength and performance rather than muscle mass [35].

Handgrip indicates muscle strength and is directly related to lower-extremity strength. Calf circumference indicates skeletal muscle mass and is associated with higher strength [36, 37]. A longitudinal cohort study conducted in community-dwelling adults (n: 633, aged: 55–65 years) analyzed the association between vitamin K status and physical functioning over 13 years. An association of low vitamin K status with lower handgrip strength, smaller calf circumference was observed. Low vitamin status in women indicated an existence of association of low vitamin status with poorer functional performance score [38].

Some observational studies conducted in sarcopenia patients showed an association of high vitamin K status in plasma with muscle strength, large muscle mass, and high physical performance.

Thus, it was concluded that physical performance scores rather than muscle mass indicated the beneficial effect of vitamin K on muscle quality [35].

Systremma or leg cramps is a common and distressing problem characterized by involuntary, painful, sudden contractions of the skeletal muscles. It has affecting 30% of people who are over sixty-year-of age and 50% of people over eighty years of age [39, 40]. Muscle cramps may occur in normal subjects during a strong voluntary contraction, sleep, sports or pregnancy but it can also occur due to several pathological conditions such as myopathies, neuropathies, motoneuron diseases, metabolic disorders, hydroelectrolyte imbalances or endocrine pathologies, cirrhosis of liver, in patients on dialysis or may be triggered by intake of certain drugs such

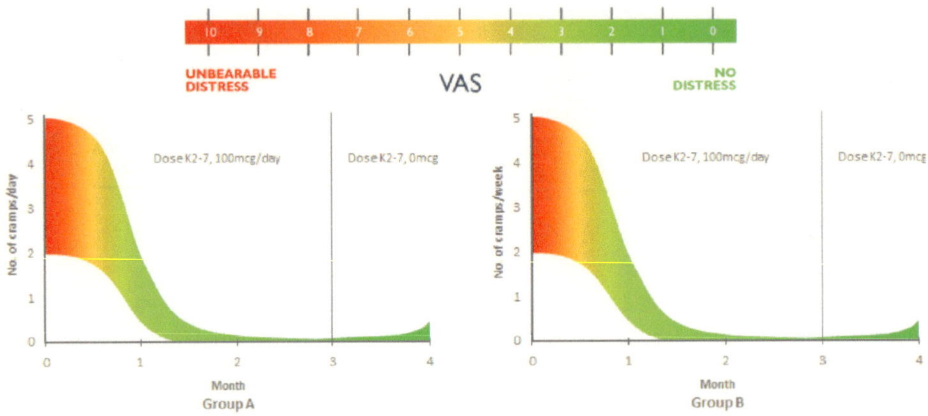

Figure 4.
Decrease in mean severity of cramps as noted on VAS score in both the groups which were divided depending upon the frequency of cramps viz., cramps every day in group A and 2–3 cramps every week in group B.

as diuretics, laxatives, beta2-agonists, cimetidine, and phenothiazines. Treatment of the underlying cause could successfully relieve this symptom [41, 42].

Diverse causes of muscle cramps has led to varied treatment modalities in clinical practice with varying degree of success in relieving the symptoms [43, 44]. These modalities include quinine Sulphate [45], calcium channel blockers [46], magnesium [47], gabapentin [48], botulinum toxin [49], phenytoin [50], Vit E [51], carisoprodol and orphenadrine [52]. Although quinine is the most used treatment modality in this condition [41], it is associated with several side effects like arrhythmia, tinnitus, headache, nausea, tremor, hypotension, and gastrointestinal upset, and occasionally, potentially fatal hypersensitivity reactions and thrombocytopenia [40, 41]. Due to severe toxicity encountered, US FDA has banned over-the-counter quinine-based products used for leg cramps [41, 53, 54]. This has generated a need for alternative therapeutic agents.

A preliminary open labeled observational study conducted by Vaidya *et al.* showed that daily administration of 100 mcg of Menaquinone-7 for 3 months was associated with a reduction in the frequency, intensity, and duration of idiopathic muscle cramps [55]. Menaquinone-7 at a dose of 100 mcg /day for 3 months was found to be well tolerated and safe and resulted in therapeutic relief of muscle cramps (**Figure 4**).

A research done by Mehta and Vaidya showed that daily administration of vitamin K relieves muscle cramps and prevents its recurrence. Vitamin MK-7 has longer half-life which facilitates its further utilization as it stays in the body for a longer duration. Vitamin K is a safe prophylactic for muscle cramps. Vitamin K also improves the muscle strength evident by relief of fatigue. The inventors have discovered relief from cramps when sufficient dose of vitamin K was administered systematically daily once or more. The preferred range was 10 μg to 1000 μg per day, and the preferred vitamin K was vitamin MK-7 [56].

6.5 Nerve health

Peripheral neuropathy (PN) also known as distal symmetric neuropathy or sensorimotor neuropathy, is a common problem with multifactorial aetiologies. Diabetes mellitus is the most common etiology of PN. Neural signals from sensory receptors in the cellular pathway of the peripheral nervous system are damaged in PN. It is a neurological complication where the nerves carrying sensory neurons from different parts of the body to the central nervous system are denervated hence causing numbness, tingling, motor paralysis and gland or organ dysfunction. PN is a disease with vast spectrum, found in a variety of groups of populations, commonly observed in geriatrics, obese and diabetic population [57].

An epidemiological study conducted by Martyn *et al.* has stated the worldwide prevalence of PN to be around 2.4% which is considerably increasing to 8% in patients older than 55 years [58].

Indian population is susceptible to PN due to large population density, exposed to different adverse environments for a living [59]. Amongst diabetic Indian population, prevalence of neuropathy has been 26–31% [60–62]. In an Indian epidemiological survey conducted amongst about 40 million diabetics in India, at least 10.4 million diabetics showed the symptoms of PN [63].

Some patients with neuropathy may experience extremely painful symptoms, whereas others may have objectively marked neurological deficit without significant painful neurological symptoms [64].

A systematic review analyzing the data of several studies stated that painful diabetic PN occurs in about one in six people with diabetes, impairing the quality of life of people and increasing healthcare cost. Although guidelines have suggested several treatment related recommendations, but they are associated with adjuvant side effects [65]. The risk of developing PN increases with the duration of diabetes and deteriorating glycemic control [64].

Neuropathy is a devastating event in patients with myeloma. Prolonged treatment of Multiple Myeloma (MM) related drugs leads to development of PN in 70% of patients [66]. Neuropathic events in such patients leads to dose reduction of the primary agents (Bortezomib, Thalidomide and Lenalidomide) or reduction in frequency of the therapy. This could further lead to discontinuation of therapy by some of the patients. Therefore, neuropathy in MM needs to be addressed.

The etiopathology of PN is poorly understood. Many factors, including dietary deficiencies, may contribute to the clinical manifestation of the condition [67].

The neurologic manifestations of folate deficiency overlap with those of vitamin B12 deficiency and include cognitive impairment, dementia, depression and commonly PN [64].

Myelopathy with or without an associated neuropathy is the commonly recognized neurological manifestations of vitamin B12 deficiency [68]. Methyl cobalamin is a vitamin B12 analogue, necessary for the maintenance of the nervous system [69]. The diagnosis of neuropathy due to B12 vitamin deficiency remains a real challenge for the clinician [70].

The etiology of diabetic neuropathy has been a debatable topic however, neuropathy due to an inflammatory autoimmune condition that damages the myelin sheet of peripheral nerves and role of menaquinone-7 deficiency in alleviating this condition are considered as the evolving possibilities for diabetic neuropathy.

Vitamin K is considered to have a role in myelin synthesis and repair in central and peripheral nervous systems. Myelin is a sphingolipid, a group of complex lipids which are found in all mammalian cells as a major components of cell membranes, present particularly in high concentrations in cells of the central and peripheral nervous systems [71]. Certain sphingolipids found in the central and peripheral nervous systems have shown a high correlation with the tissue levels of vitamin K. Initially recognized for their structural role, sphingolipids are now considered as the key players in major cellular events such as proliferation, differentiation, senescence, cell–cell interaction, and transformation [72]. Furthermore, several recent research have shown a corelation of alterations in sphingolipids metabolism with aging process [73] and neurodegenerative disorders such as Alzheimer's disease (AD) and Parkinson's disease [52, 74].

Scientific legacy of Meir Lev's group depicted the role for vitamin K in sphingolipid metabolism in a report published in Nature in 1958 [75]. The report showed that Vitamin K serves as a growth factor for the rumen strain *Bacteroides melaninogenicus* (also known as *Fusiformis nigrescens*), which was later found to be linked to cell membrane homeostasis. When *Bacteroides melaninogenicus* was cultured in a medium without vitamin K, cells grew as filaments (i.e., elongated cells), were more fragile when subjected to shaking with glass beads and tended to auto-agglutinate when placed in buffer. Growth of bacteria under such condition was also greatly affected. Vitamin K deficient cultures yielding around 80% lower bacteria weight than those grown from vitamin K replete conditions [76]. Two other reports also explained the role of vitamin K at the membrane level. The reports showed the essential requirement of vitamin K for sphingolipid synthesis. Recent published studies that confirm the modulation of brain sphingolipids by vitamin K nutritional status, underscore the potentially far-reaching effect of vitamin K in brain function given the key role of these lipids in cell-signaling functions.

In the nervous system, vitamin K activates the carboxylation and activation of Gla residues on GAS6 protein (growth arrest-specific gene 6 protein) which is structurally related to another vitamin K-dependent protein (VKDP), anticoagulation factor protein S [77]. GAS6 and related S protein bind and activate the receptor tyrosine kinases of the Tyro3, Axl, and Mer (TAM) family. They are responsible for cell signaling which stimulates the generation of central nervous system repair cells (oligodendrocytes) and increased myelin production including repair after myelin injury (demyelinating injury) [78]. Vitamin K may also act in the central nervous system independent to its role in the carboxylation reaction [79]. Vitamin K independent of VKDP, activates enzyme 3-ketodihydrosphingosine (3-KDS), involved in sphingolipid synthesis which is critical for healthy myelin [80].

Sakaue M et al. investigated the protective effects of different forms of Vitamin K (Vitamin K1 and Vitamin K2–4) in an *in vitro* experiment conducted in primary cultured neurons from cerebella of rat pups where methylmercury-induced the cell death. They also investigated its protective effect against GSH-depletion-induced cell death by employing two intracellular glutathione (GSH) reducers, L-buthionine sulfoximine (BSO) and diethyl maleate (DEM), in primary cultured neurons and human neuroblastoma IMR-32 cells. It was observed that all the forms of Vitamin K inhibited the death of the primary cultured neurons indicating that vitamin K forms have the potential to protect neurons against cytotoxic methylmercury and agents that deplete GSH, without increasing intracellular GSH levels [81]. Kenji Onodera et al. while examining the antinociceptive effects of Vitamin K2–4 in diabetic mice found that no significant difference exist between non-diabetic and diabetic mice in the Vitamin K2–4 induced changes in the nociceptive threshold. This indicated the therapeutic effectiveness of Vitamin K2–4 for treating painful diabetic neuropathy [82].

A serendipitous discovery by two researchers, Mehta and Vaidya is that Menaquinone-7 relieves idiopathic muscle cramps as well as symptoms of diabetic neuropathy. PCT/IN2008/000465, application further claims the safety of usage of Menaquinone-7 in the various novel conditions like neuropathy [56].

In an open labeled study conducted by Kulkarni et al., it was shown that Menaquinone-7 at a dose of 100 mcg twice a day for 8 weeks (**Figure 5**) was well tolerated and safe with a therapeutic activity for the symptoms of peripheral neuropathy [83].

Based on the results of these studies, the next study which is a follow-up study in a larger cohort (n = 100) was planned to address the peripheral neuropathy experienced by patients. Menaquinone-7 capsules (100 mcg / capsule, twice a day) were given orally for 8 weeks and were followed up to 12 weeks. By twelfth week, the score was reduced in megaloblastic anemia as well as in diabetes mellitus groups to

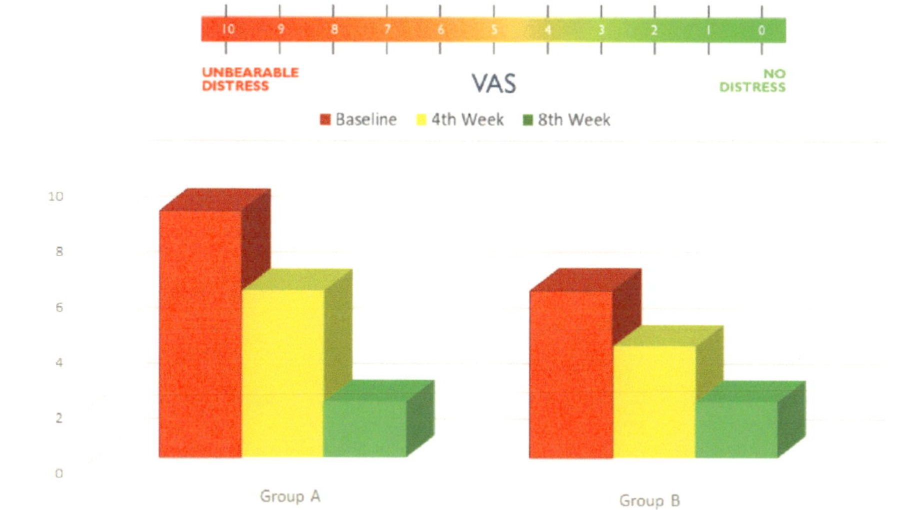

Figure 5.
Decrease in the intensity and severity of PN from baseline to 8th week as noted on VAS score in the groups a and B, where group A (severe) had a VAS score of 8–9 and group B (moderate) with a score of 6–8 at baseline.

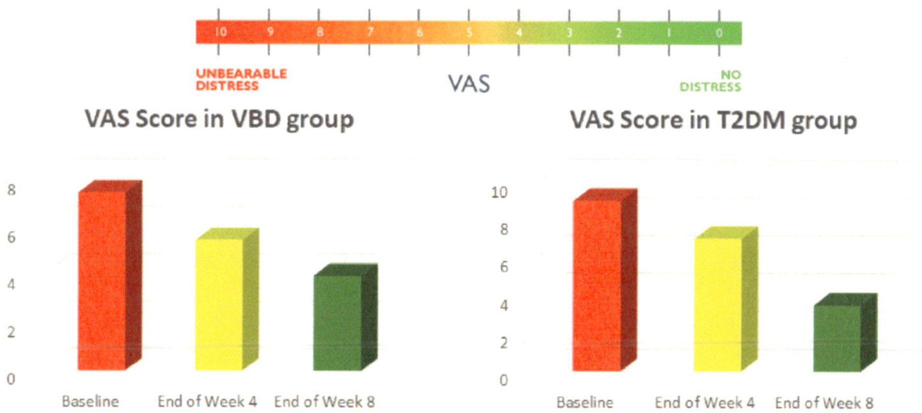

Figure 6.
Decrease in the intensity and severity of PN from baseline to 8th week as noted on VAS score in the groups VBD (Vitamin B12 deficiency) and T2DM (type 2 diabetes mellitus).

1–2 (**Figure 6**). The decrease was statistically significant ($P < 0.0001$). The tingling and numbness had reduced significantly. There was a significant decrease in the weakness and fatigue [84].

Recently a double-blind placebo-controlled efficacy and safety study of Menaquinone-7 was conducted in 60 patients presenting with peripheral neuropathy and suffering from either vitamin B12 deficiency and/ or type 2 diabetes mellitus. Patients from both the groups' i.e., Vitamin B12 deficiency and type 2 diabetes mellitus had overall VAS score of 9 at baseline. By the end of the twelfth week, patients who were receiving Menaquinone-7 showed statistically significant reduction in the VAS score in Vitamin B12 deficiency as well as in type 2 diabetes mellitus to 2; whereas the patients who were taking placebo in Vitamin B12 deficiency group had reduced to 8, and in type 2 diabetes mellitus group to 9. This study was again performed with same protocol along with estimation of serum Menaquinone-7 levels in serum in a small sample size. The VAS score showed an inverse relationship between Menaquinone-7 levels and peripheral neuropathy symptoms (**Figure 7**) [85, 86].

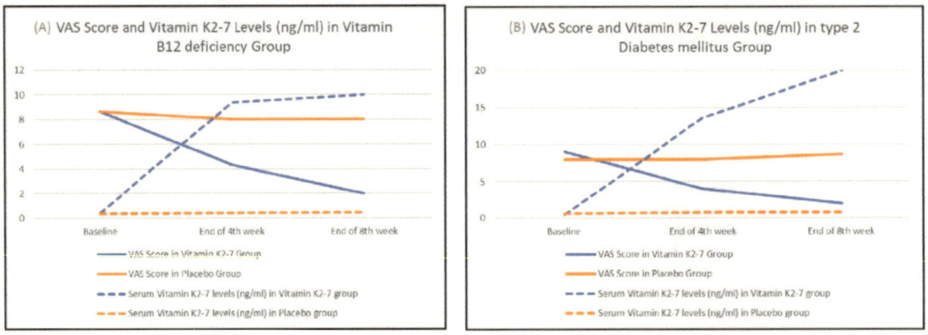

Figure 7.
Average VAS score and serum Vitamin K2–7 levels (ng/ml) in Vitamin B12 deficiency group (A) and type 2 diabetes mellitus group (B).

Antineoplastic agents are the chemotherapy drugs used for cancer. Chemotherapy-induced peripheral neuropathy (CIPN) is one of the most frequent side effects caused by antineoplastic agents having a prevalence ranging from 19% to over 85%. CIPN is a mostly sensory neuropathy and is associated with motor and autonomic changes of varying intensity and duration [87]. Chemotherapeutics induces toxicity in peripheral nervous system. Oxaliplatin, an antineoplastic agent damages the blood brain barrier (BBB). The possible mechanisms of BBB damage may include proinflammatory cytokines, ROS, or other neurotransmitters, all of which are involved in the peripheral nervous system toxicity induced by chemotherapeutics [88, 89]. The study of Sanna et al. has shown a direct correlation between structural changes in the central nervous system and chemotherapy-induced neurotoxicity [90].

An open labeled observational study to evaluate the iatrogenic neuropathy and its amelioration using Menaquinone-7 in patients with Multiple Myeloma with drug induced PN suggests for the first time that Menaquinone-7 has an ameliorative potential for relief of iatrogenic PN in Multiple Myeloma patients. Menaquinone-7 reduces the symptoms of PN like tingling, numbness, burning sensation, pain, causalgia, wooly feeling, and cramps caused during treatment of MM, thus Menaquinone-7 is found to be useful in the treatment of PN caused due to the therapy of MM [91].

Multiple myeloma (MM) is a type of hematological cancer which is characterized by excessive production of malignant plasma cell clones in the bone marrow [92]. Incidence of iatrogenic PN has been observed in patients with MM who received chemotherapy. It is primarily of a sensory or sensorimotor nature, and the symptoms of tingling, numbness, burning sensation and pain are predominantly bilaterally symmetric [93]. Development of debilitating drug induced PN is one of the major challenges in the treatment of MM, affecting compliance leading to discontinuation of therapy or dose/drug modification [94]. Thus, there is a need of any modality that could reduce the severity and allows continuation of effective therapy in the clinical setting. This preliminary observational study is the first study revealing the potential of Menaquinone-7 in relieving the symptoms of iatrogenic PN in MM patients.

7. Conclusion

Menaquinone-7 appears promising in the areas of chronic degenerative conditions such as bone health, cardiovascular, diabetes, energy metabolism, peripheral neuropathy, cramps etc. Newer research is ongoing to confirm its role in many other areas including immunity, cognition, cancer etc. With the recent discovery of its many biological functions, Menaquinone-7 is sometimes referred to as a multitasking vitamin. Globally speaking Menaquinone-7 is negligible in diet consumed by the population all around the world except in small pockets leading to Menaquinone-7 insufficiency. Either lack or deficiency of a given vitamin invites multiple pathologies, some mild some severe, some experiential some silent, some acute some chronic. Until this knowledge is available to an individual, he is likely to consider multiple healing effects of a vitamin as panacea. Researchers have now realized that most of the global population is facing multiple severe morbidities due to lack of or inadequate levels of Menaquinone-7 in diet and supplementation. The "Next Big Thing" in medicine is Menaquinone-7 with what the science has revealed already. This vitamin will advance on an exponential curve.

Conflict of interest

Dr. Dilip S. Mehta, Dr. Anselm de Souza, and Dr. Shashank S. Jadhav are CEO, Managing Director and Medical Director of Synergia Life Sciences Pvt. Ltd.

Author details

Dilip Mehta, Anselm de Souza and Shashank S. Jadhav*
Synergia Life Sciences Pvt. Ltd., Mumbai, India

*Address all correspondence to: shashank@viridisbiopharma.com

References

[1] Shampo, M.A., Henrik Dam Kyle, R.A., 1998. Discoverer of Vitamin K. Shampo, Marc A. et al. Mayo Clinic Proceedings, Volume 73, Issue 1, 46.

[2] Schurgers L.J., Vermeer C. Determination of phylloquinone and menaquinones in food. Effect of food matrix on circulating vitamin K concentrations. Haemostasis. 2000;30:298-307.

[3] Ravishankar B, Dound YA, Mehta DS, Ashok BK, de Souza A, Pan MH, *et al*. Safety assessment of menaquinone-7 for use in human nutrition. Journal of Food and Drug Analysis 2015;23:99-108.

[4] Pucaj K, Rasmussen H, Moller M, Preston T. Safety and toxicological evaluation of a synthetic vitamin K2, menaquinone-7. Toxicology Mechanisms and Methods 21(7):520-32

[5] EFSA. 2008. Scientific Opinion. Vitamin K2 added for nutritional purposes in foods for particular nutritional uses, food supplements and foods intended for the general population. The EFSA Journal 822:1-31.

[6] Marles RJ, Roe AL, Oketch-Rabah HA. 2017. US Pharmacopeial Convention safety evaluation of menaquinone-7, a form of vitamin K. Nutr Rev. 75(7):553-578.

[7] IOM. 2000. Dietary reference intakes for vitamin A, vitamin K, arsenic, boron, chromium, copper, iodine, iron, manganese, molybdenum, nickel, silicon, vanadium and zinc. National Academy of Sciences. Institute of Medicine Food and Nutrition Board. pp. 162-196.

[8] Sakano T, Notsumoto S, Nagaoka T, Morimoto A, Fujimoto K, Masuda S. Measurement of K vitamins in food by high-performance liquid chromatography with fluorometric detection. Vitamins. 1988; 62: 393-398.

[9] Vermeer C, Raes J, Hoofd C, Knapen MHJ, Xanthoulea S. Menaquinone Content of Cheese. Nutrients. 2018 Apr; 10(4): 446.

[10] Kamao, M., Suhara, Y., Tsugawa, N., Uwano, M., Yamaguchi, N., Uenishi, K., Ishida, H., Sasaki, S., Okano, T., 2007. Vitamin K content of foods and dietary vitamin k intake in Japanese young women. J. Nutr. Sci. Vitaminol., 53, 464-470.

[11] Kaneki M, Stephen J. Hedges, Takayuki Hosoi *et al.* (2001) Japanese fermented soybean food as the major determinant of the large geographic difference in circulating levels of vitamin K2: Possible implications for hip-fracture risk. Nutrition. 17, 315-321.

[12] Shearer, M. J., Bach, A. & Kohlmeier, M. (1996) Chemistry, nutritional sources, tissue distribution and metabolism of vitamin K with special reference to bone health. J. Nutr. 126 (suppl.): 1181S–1186S.

[13] Suttie, J. W. (1995). The importance of menaquinones in human nutrition. Annu. Rev. Nutr. 15: 399-417.

[14] Karl JP, Meydani M, Barnett JB, Vanegas SM, Barger K, Fu X, Goldin B, Kane A, Rasmussen H, Vangay P, Knights D, Jonnalagadda SS, Saltzman E, Roberts SB, Meydani SN, Booth SL. Fecal concentrations of bacterially derived vitamin K forms are associated with gut microbiota composition but not plasma or fecal cytokine concentrations in healthy adults. Am J Clin Nutr. 2017 Oct;106(4):1052-1061.

[15] Geleijnse JM, Vermeer C, Grobbee DE, Schurgers LJ, Knapen MH, van der Meer IM, et al. Dietary Intake of

Menaquinone Is Associated with a Reduced Risk of Coronary Heart Disease: The Rotterdam Study. The Journal of Nutrition 2004;134: 3100-3105.

[16] Gast GC, de Roos NM, Sluijs I, Bots ML, Beulens JW, Geleijnse JM, et al. A high menaquinone intake reduces the incidence of coronary heart disease. Nutr Metab Cardiovasc Dis 2009;19:504-510.

[17] Knapen, M.H., Drummen, N.E., Smit, E., Vermeer, C., Theuwissen, E., 2013. Three-year low-dose menaquinone-7 supplementation helps decrease bone loss in healthy postmenopausal women. Osteoporosis International. Osteoporos Int. 2013 Sep;24(9):2499-2507.

[18] Sofie Hertz Rønn, Torben Harsløf, Steen Bønløkke Pedersen and Bente Lomholt Langdahl. Vitamin K2 (menaquinone-7) prevents age-related deterioration of trabecular bone microarchitecture at the tibia in postmenopausal women. European Journal of Endocrinology (2016) 175, 541-549.

[19] S. Kanellakis et al. Changes in Parameters of Bone Metabolism in Postmenopausal Women Following a 12-Month Intervention Period Using Dairy Products Enriched with Calcium, Vitamin D, and Phylloquinone (Vitamin K1) or Menaquinone-7 (Vitamin K2): The Postmenopausal Health Study II. Calcif Tissue Int (2012) 90:251-262.

[20] van Summeren M, Braam L, Noirt F, Kuis W, Vermeer C. Pronounced elevation of undercarboxylated osteocalcin in healthy children. Pediatr Res. 2007 Mar;61(3):366-370.

[21] Spronk HMH., Soute BAM, Schurgers LJ, Thijssen HHW, De Mey JGR, Vermeer C. Tissue-Specific Utilization of Menaquinone-4 Results in the Prevention of Arterial Calcification in Warfarin-Treated Rats. J Vasc Res 2003;40:531-537.

[22] Yamaguchi M, Sugimoto E and Hachiya S. Stimulatory effect of menaquinone-7 (vitamin K2) on osteoblastic bone formation in vitro. Mol Cell Biochem. 2001 Jul;223(1-2):131-137.

[23] Min Zhu, Jinxing Ma, Sheng Lu, Yueliang Zhu, Yi Cui, Hongbo Tan et al. Vitamin K2 analog menaquinone-7 shows osteoblastic bone formation activity in vitro. Biomedical Research 2017; 28 (3): 1364-1369.

[24] Ewa Ambrozewicz, Marta Muszynska, Grażyna Tokajuk, Grzegorz Grynkiewicz, Neven Zarkovic and Elzbieta Skrzydlewska. Beneficial Effects of Vitamins K and D3 on Redox Balance of Human Osteoblasts Cultured with Hydroxyapatite-Based Biomaterials. Cells 2019, 8(4), 325.

[25] Beulens JW, van der A DL, Grobbee DE, Sluijs I, Spijkerman AM, van der Schouw YT. Dietary Phylloquinone and Menaquinones Intakes and Risk of Type 2 Diabetes. Diabetes Care 2010;33:1699-1705.

[26] Choi HJ, Yu J, Choi H, An JH, Kim SW, Park KS, *et al*. Vitamin K2 Supplementation Improves Insulin Sensitivity via Osteocalcin Metabolism: A Placebo-Controlled Trial. Diabetes Care 2011;34:e147.

[27] M Yoshida, PF Jacques, JB Meigs, et al. Effect of vitamin K supplementation on insulin resistance in older men and women. Diabetes Care. 2008 Nov;31(11):2092-2096.

[28] Brian K. McFarlin, Andrea L. Henning, Adam S. Venable. Oral Consumption of Vitamin K2 for 8 Weeks Associated with Increased Maximal Cardiac Output During Exercise. Altern Ther Health Med. 2017;23(4):26-32.

[29] Vos M, Esposito G, Edirisinghe JN, Vilain S, Haddad DM, Slabbaert JR, et al. Vitamin K2 is a mitochondrial electron carrier that rescues pink1 deficiency. Science 2012;336, 1306-1310.

[30] Prasuhn J, Kasten M, Vos M, König IR, Schmid SM, Wilms B, Klein C and Brüggemann N (2021) The Use of Vitamin K2 in Patients With Parkinson's Disease and Mitochondrial Dysfunction (PD-K2): A Theranostic Pilot Study in a Placebo-Controlled Parallel Group Design. Front. Neurol. 2014;11:59.

[31] Kleinman RE, Fracchia MS. Vitamin K and cystic fibrosis: give me a double, please. Am J Clin Nutr. 2010;92(3):469-470.

[32] Iwamoto J. Vitamin K_2 therapy for postmenopausal osteoporosis. Nutrients. 2014;6(5):1971-1980.

[33] Pan MH, Maresz K, Lee PS, Wu JC, Ho CT, Popko J, Mehta DS, Stohs SJ, Badmaev V. Inhibition of TNF-α, IL-1α, and IL-1β by Pretreatment of Human Monocyte-Derived Macrophages with Menaquinone-7 and Cell Activation with TLR Agonists In Vitro. J Med Food. 2016 Jul;19(7):663-669.

[34] Chatrou ML, Reutelingsperger CP, Schurgers LJ, et al. Role of vitamin K-dependent proteins in the arterial vessel wall, Hamostaseologie. 2011;31:251-257.

[35] Azuma K, Inoue S. Multiple Modes of Vitamin K Actions in Aging-Related Musculoskeletal Disorders. Int. J. Mol. Sci. 2019, 20, 2844

[36] Jakobsen LH, Rask IK, Kondrup J, et al. Validation of handgrip strength and endurance as a measure of physical function and quality of life in healthy subjects and patients, Nutrition. 2010; 26:542-550.

[37] Rolland Y, Lauwers-Cances V, Cournot M, et al. Sarcopenia, calf circumference, and physical function of elderly women: a cross-sectional study. J. Am. Geriatr. Soc. 2003; 51:1120-1124.

[38] Ballegooijena AJ, Puttena SR, Vissera M, et al. Vitamin K status and physical decline in older adults—The Longitudinal Aging Study Amsterdam. Maturitas. 2018; 113: 73-79.

[39] Naylor JR, Young JB: A general population survey of rest cramps. Age Ageing, 1994; 23: 418-420

[40] Abdulla A, Jones PW, Pearce VR: Leg cramps in the elderly: prevalence, drug and disease associations. IJCP, 1999; 53(7): 494-496

[41] McGee SR: Muscle cramps. Arch Intern Med, 1990; 150: 511-518

[42] Riley JD, Antony SJ: Leg cramps. Differential diagnosis and management. American Family Physician, 1995; 52: 1794-1798.

[43] Riley JD, Antony SJ: Leg cramps. Differential diagnosis and management. American Family Physician, 1995; 52: 1794-1798

[44] Coppin RJ, Wicke DM and Little PS. Managing nocturnal leg cramps — calf-stretching exercises and cessation of quinine treatment: a factorial randomised controlled trial. British Journal of General Practice 2005; 55: 186-191.

[45] Woodfield R, Goodyear-Smith F and Arroll B, N-of-1 trials of quinine efficacy in skeletal muscle cramps of the leg. British Journal of General Practice 2005; 55: 181-185

[46] Peer G, Blum M, Aviram A. Relief of hemodialysis-induced muscular cramps by nifedipine. Dialysis and Transplantation 1983; 12:180-181

[47] Roffe C, Sills S, Crome P, Jones P. Randomised, cross-over, placebo controlled trial of magnesium citrate in the treatment of chronic persistent leg

cramps. Med Sci Monit. 2002 May; 8(5):CR326-CR330.

[48] Serrao, M; Rossi, P ; Cardinali, P; Valente, G; Parisi, L; Pierelli, Francesco. Gabapentin Treatment for Muscle Cramps: An Open-Label Trial. Clinical Neuropharmacology. 2000;23(1):45-49.

[49] Bertolasi L, Priori A, Tomelleri G, Bongiovanni LG, Fincati E, Simonati A, De Grandis D, Rizzuto N. Botulinum toxin treatment of muscle cramps: a clinical and neurophysiological study. Ann Neurol. 1997 Feb; 41(2):181-186

[50] Minaker KL, Flier JS, Landsberg L, Young JB, Moxley RT, Kingston WJ, Meneilly GS, Rowe JW. Phenytoin-induced improvement in muscle cramping and insulin action in three patients with the syndrome of insulin resistance, acanthosis nigricans, and acral hypertrophy. Arch Neurol. 1989 Sep; 46(9):981-985

[51] Connolly PS, Shirley EA, Wasson JH, Nierenberg DW. Treatment of nocturnal leg cramps. A crossover trial of quinine vs vitamin E. Arch Intern Med. 1992 Sep;152(9):1877-1880

[52] Guay DR. Are there alternatives to the use of quinine to treat nocturnal leg cramps? Consult Pharm. 2008 Feb; 23(2):141-156

[53] Leclerc KM, Landry FJ. Benign nocturnal leg cramps. Current controversies over use of quinine. Postgrad Med. 1996 Feb;99(2):177-178, 181-4.

[54] US Department of Health and Human Services. Drug products for the treatment and/or prevention of nocturnal leg muscle cramps for over-the-counter human use. Federal Register 1994; 59: 43234-43252.

[55] Mehta DS, Vaidya RA, Dound YA, Nabar NS, Pandey SN, Vaidya ADB. Therapeutic activity and safety of Menaquinone-7 in muscle cramp: An interventional case-series. The Indian Practitioner 2010;63(5):287-297.

[56] Onodera K, Zushida K, Kamei J. Antinociceptive effect of vitamin K2 (menatetrenone) in diabetic mice. Jpn J Pharmacol. 2001 Mar;85(3):335-337.

[57] Hilton T N, Tuttle LJ, Bohnert KL, Mueller MJ, Sinacore DR. Excessive adipose tissue infiltration in skeletal muscles in individuals with obesity, diabetes mellitus, and peripheral neuropathy: Association with performance and function. Physical Therapy 2008;88:1336-1344.

[58] Martyn CN, Hughes RA. Epidemiology of peripheral neuropathy. J Neurol Neurosurg Psychiatry 1997;62:310-318.

[59] Trivedi S, Pandit A, Ganguly G, Das SK. Epidemiology of peripheral neuropathy: An Indian perspective. Ann Indian Acad Neurol 2017;20:173-184.

[60] Pradeepa R, Rema M, Vignesh J, Deepa M, Deepa R, Mohan V. Prevalence and risk factors for diabetic neuropathy in an urban south Indian population: The Chennai Urban Rural Epidemiology Study (CURES-55). Diabet Med 2008;25:407-402.

[61] Paul UK, Sinharay K, Bhattacharyya AK, Pal SK. Update in diabetic neuropathy with special reference to Indian scenario. J Indian Med Assoc 2012;110:616-622.

[62] Gill H K, Yadav S B, Ramesh V, Bhatia E. A prospective study of prevalence and association of peripheral neuropathy in Indian patients with newly diagnosed type 2 diabetes mellitus. J Postgrad Med 2014;60:270-275.

[63] Shashank R. Joshi. Diabetes Care in India. Annals of Global Health. Vol. 81, No. 6, 2015.

[64] Paisley P and Serpell M. Improving pain control in diabetic neuropathy. Practitioner. 2017 Mar;261(1802):23-26

[65] Rosenberg CJ and Watson JC. Treatment of painful diabetic peripheral

neuropathy. Prosthet Orthot Int. 2015 Feb;39(1):17-28.

[66] Martin TG. Peripheral neuropathy experience in patients with relapsed and/or refractory multiple myeloma treated with carfilzomib. Oncology 2013;27:4-10.

[67] Franques J, Gazzola S. Metabolic and nutritional neuropathies: update in diabetes, vitamin B12 and copper deficiency. Rev Neurol (Paris). 2013 Dec; 169(12):991-996.

[68] Reynolds EH. The neurology of folic acid deficiency. Handb Clin Neurol. 2014; 120: 927-943.

[69] Kumar N. neurologic aspects of cobalamin (B12) deficiency. Handb Clin Neurol. 2014; 120: 915-926.

[70] Tanaka H. Old or new medicine? Vitamin B12 and peripheral nerve neuropathy. Brain Nerve. 2013 Sep;65(9):1077-1082.

[71] Carrie I, Portoukalian J, Vicaretti R, Rochford J, Potvin S, Ferland GJ Nutr. 2004 Jan; 134(1):167-172.

[72] Zeidan YH, Hannun YA. Translational aspects of sphingolipid metabolism. Trends Mol Med. 2007;13:327-336

[73] Cutler RG, Kelly J, Storie K, Pedersen WA, Tammara A, Hatanpaa K, Troncoso JC, Mattson MP. Involvement of oxidative stress-induced abnormalities in ceramide and cholesterol metabolism in brain aging and Alzheimer's disease. Proc Natl Acad Sci U S A. 2004;101:2070-2075.

[74] Posse de Chaves E, Sipione S. Sphingolipids and gangliosides of the nervous system in membrane function and dysfunction. FEBS Lett. 2010; 584:1748-59.

[75] Lev M. Apparent requirement for vitamin K of rumen strains of Fusiformis nigrescens. Nature. 1958;181:203-204.

[76] Lev M. Vitamin K deficiency in Fusiformis nigrescens. I. Influence on whole cells and cell envelope characteristics. J Bacteriol. 1968;95:2317-2324.

[77] Manfioletti G, Brancolini C, Avanzi G, Schneider C Mol Cell Biol. 1993 Aug; 13(8):4976-4985.

[78] Ferland G. "Vitamin K and the Nervous System: An Overview of its Actions" Adv. Nutr. 2012, 3, 204-212.

[79] Lev M, Sundaram S. Modulation of glycosphingolipid synthesis by vitamin K depletion in bacteria and brain. In: Suttie JW, editor. Current Advances in Vitamin K Research: Proceedings of the 17th Steenbock Symposium. Madison: University of Wisconsin-Madison; 1988.

[80] Lev M, Milford AF Arch Biochem Biophys. 1973 Aug; 157(2):500-508.

[81] Sakaue M, Mori N, Okazaki M *et al* Vitamin K has the potential to protect neurons from methylmercury-induced cell death in vitro. J Neurosci Res. 2011 Jul;89(7):1052-1058.

[82] Vaidya AB, Mehta DS, de' Souza A, Vaidya RA, inventors. Viridis BioPharma Pvt. Ltd., assignee. Treatment using Vitamin K Analogues and Derivatives. PCT Patent, 2009. WO 2009/063485 A2.

[83] Kulkarni VK, Upase DP, Dound YA, Jadhav SS, Bhave AA, Mehta DS et al. The effect of vitamin B12 deficiency and/or diabetes mellitus: A preliminary study. The Indian Practitioner 2013;66(10):625-629.

[84] Mehta DS, Dound YA, Jadhav SS, Bhave AA, Devale Milind, Vaidya ADB. A novel potential role of Menaquinone-7 in relieving peripheral neuropathy. The Indian Practitioner 2019;9(4):180-185.

[85] Mehta DS, Jadhav SS, Devale M, et al. Randomized Controlled Trial of the Antinociceptive Effect of Menaquinone-7 (MK-7) in Patients with Peripheral Neuropathy. EC Neurology 13.8 (2021): 01-10.

[86] Mehta DS, Jadhav SS, Devale M, et al. Randomized, Placebo-Controlled, Double-Blind Pilot Study of the Antinociceptive Effect of Menaquinone-7 (Vitamin MK-7) in Patients with Peripheral Neuropathy. EC Neurology 13.9 (2021): 112-124.

[87] Renata Za, Kocot-Kępska M, Leppert W, et al. Mechanisms of Chemotherapy-Induced Peripheral Neuropathy. Int. J. Mol. Sci. 2019; 20: 1451. doi:10.3390/ijms20061451

[88] Rochfort K.D, Collins L.E, Murphy R.P, et al. Downregulation of blood–brain barrier phenotype by proinflammatory cytokines involves NADPH oxidase-dependent ROS generation: Consequences for interendothelial adherens and tight junctions. PLoS ONE 2014, 9, e101815.

[89] Argaw A.T, Zhang Y, Snyder B.J, et al. IL-1beta regulates blood–brain barrier permeability via reactivation of the hypoxia angiogenesis program. J. Immunol. 2006, 177, 5574-5584.

[90] Sanna M.D, Ghelardini C, Galeotti N. Altered Expression of Cytoskeletal and Axonal Proteins in Oxaliplatin-Induced Neuropathy. Pharmacology 2016, 97, 146-150.

[91] Bhave AA, Mehta DS, Dound YA, Jadhav SS, Vaidya RA, Vaidya ADB. Menaquinone-7, a promising solution to peripheral neuropathy aroused due to multiple myeloma treatment. The Indian Practitioner 2019;72(1):16-20.

[92] Dimopoulos MA and Terpos E. Multiple myeloma. *Ann Oncol. 2010*;21: vii143-50.

[93] Lacy MQ, Hayman SR, Gertz MA, et al. Pomalidomide (CC4047) plus low-dose dexamethasone (pom/dex) is highly effective therapy in relapsed multiple myeloma. J Clin Oncol 2009;27:5008-5014.

[94] Terpos E, Suzan F, Goldschmidt H. Going the distance: Are we losing patients along the multiple myeloma treatment pathway? Critical Reviews in Oncology/Hematology. 2018;126: 19-23.

Chapter 5

Overcoming the Photochemical Problem of Vitamin K in Topical Application

Shotaro Goto, Shuichi Setoguchi, Kazuhisa Matsunaga and Jiro Takata

Abstract

Topical application of vitamin K is beneficial in the treatment of various skin pathologies. However, its delivery to the skin is hampered by the photo-instability and phototoxicity of vitamin K (quinone form). Indeed, topical use of vitamin K is regulated in Europe owing to the photosensitive properties of this molecule. Here, we discuss the suitability of ester derivatives of vitamin K hydroquinone (VKH), the active form of vitamin K, for topical applications. Notably, VKH derivatives have the potential to overcome the photo-instability and phototoxicity problem of vitamin K and act as VKH prodrugs, as demonstrated in HaCaT human keratinocytes. Thus, VKH prodrug is a promising strategy for topical application of vitamin K without the need for special protection from light.

Keywords: vitamin K, photostability, phototoxicity, skin application, prodrug

1. Introduction

Skin application of vitamin K shows several beneficial effects, such as suppression of pigmentation and alleviation of bruising [1–3], prophylactically limiting the occurrence of acneiform side effects in patients receiving the monoclonal antibody cetuximab [4–6] and promoting wound healing [7].

Despite these potentially beneficial effects, vitamin K is unstable in the presence of light [8, 9]. Indeed, application of vitamin K on the skin can result in photodegradation without appropriate shielding of the application site, e.g., the face and hands, from light. Furthermore, in Europe, warnings have been issued regarding the use of vitamin K in cosmetics. The Scientific Committee on Consumer Safety has also reported phototoxicity of vitamin K in skin cells [10, 11]. As a result, the use of vitamin K as an external preparation for the skin is limited.

The molecules in the vitamin K family contain 2-methyl-1,4-naphthoquinone as the basic skeleton. Vitamin K molecules can be classified as phylloquinone (PK, vitamin K1) with a phytyl side chain at the 3-position, menaquinone (MK-n, vitamin K2) with an isoprenyl side chain consisting of n isoprenyl groups, and menadione (MD, vitamin K3) with no side chain at the 3-position. PK and MK-4 are widely used as pharmaceutical treatments for vitamin K deficiency and osteoporosis.

Vitamin K (quinone form) delivered intracellularly is converted into vitamin K hydroquinone (VKH) by two-electron reduction. VKH functions as a cofactor for

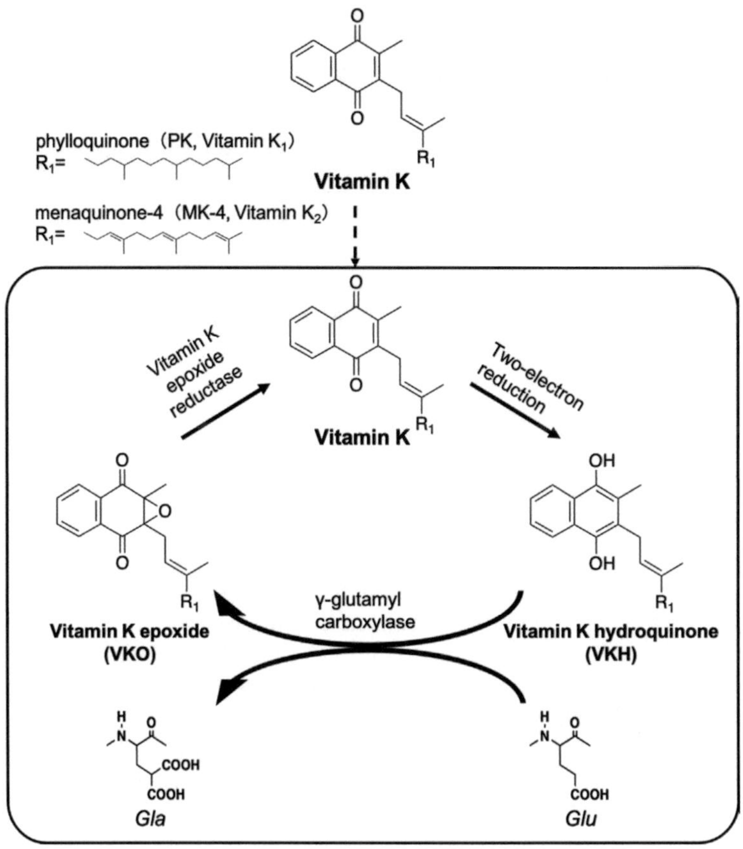

Figure 1.
Schematic illustration of the vitamin K cycle.

γ-glutamyl carboxylase (GGCX), which converts the glutamic acid (Glu) residue of vitamin K-dependent protein into the γ-carboxyglutamic acid (Gla) residue as a post-translational modification. Subsequently, VKH is oxidized to vitamin K epoxide (VKO). In addition, VKO is reduced to vitamin K (quinone form) to form the vitamin K cycle. Therefore, it is necessary to deliver sufficient VKH to the target site to achieve efficacy.

The active forms of PK and MK-4 are phyllohydroquinone (PKH) and menahydroquinone-4 (MKH), respectively. However, these compounds cannot be used as preparations because they show extreme instability via oxidation. Therefore, PK and MK-4 that are stable against oxidation are used clinically. However, as described above, vitamin K (quinone form) is extremely unstable upon exposure to light. Thus, strict control of lighting is required during formulation, distribution, storage at medical institutions, and administration to patients. To achieve full efficacy of vitamin K, formulations with low photodegradation and phototoxicity are needed for effective topical delivery of VKH. The concepts underlying the delivery system of VKH using Vitamin K are shown in **Figure 1**.

2. Mechanisms of photodegradation and phototoxicity of vitamin K

Analysis of the PK photolysis reaction by Hangarter et al. showed that charge transfer from the β,γ-double bond of the isoprenyl side chain to the quinone moiety initiates intramolecular proton transfer from the side chain and yields a 1,3-quinone methide (meta-quinone methide) as a mixture of singlet and triplet species diradical in polar solvents, subsequently forming 1,2-quinone methide (ortho-quinone methide), which can be used to generate PK chromenol [9]. Chromenol levels tend to increase with irradiation time, and this compound is expected to be the final product of photodegradation of vitamin K [9, 12, 13].

Vitamin K is also expected to cause two types of phototoxicity during the abovementioned series of photodegradation processes. After acquiring the excited state via light absorption, some chemicals cause oxidative damage to biological components, such DNA and proteins, through the generation of free radicals (type I reaction) by the electron rearrangement reaction and the generation of singlet oxygen from triplet oxygen (ground state) by the energy rearrangement reaction (type II reaction) [14, 15].

We have previously confirmed that irradiation of PK and MK-4 with UVA increases singlet oxygen generation, intracellular reactive oxygen species (ROS) generation, and cytotoxicity in HaCaT human keratinocytes. Thus, vitamin K has phototoxic properties [12, 13]. Moreover, 1,3-quinone methide diradical and 1,2-quinone methide, which are produced during the photodegradation of vitamin K, are highly reactive and show phototoxicity via type I reactions. Additionally, singlet oxygen generation through a type II reaction is an early-stage phototoxic reaction that generates additional secondary ROS [16]. Singlet oxygen is also known to promote the peroxidation of skin surface lipids, resulting in the induction of skin inflammation [17]. Because no enzymes are known to scavenge singlet oxygen in the body, singlet oxygen is thought to exhibit extremely strong cytotoxicity. **Figure 2** shows a schematic diagram of the processes of vitamin K photodegradation and phototoxicity.

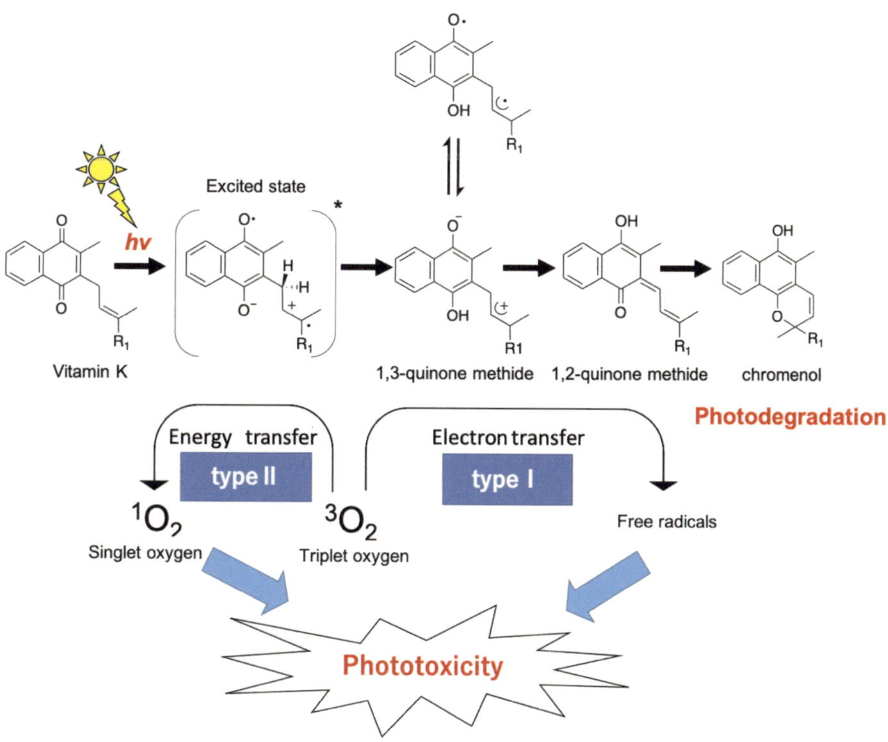

Figure 2.
Schematic diagram of vitamin K photodegradation and phototoxicity.

3. Application of VKH derivatives to overcome vitamin K photodegradation and phototoxicity

3.1 VKH derivatives strategy for applying vitamin K to the skin

Vitamin K photodegradation and phototoxicity are derived from its quinone structure [9]. Therefore, as far as it has a quinone structure, the photoreaction is unavoidable. We have previously synthesized VKH derivatives without a quinone structure in which the two hydroxyl groups of the VKH are protected by ester bonds. In addition, we have previously reported cationic VKH derivatives which have *N*,*N*-dimethylglycine (DMG) and anionic VKH derivatives which have succinic acid (SUC) as promoieties (**Figure 3**) can deliver VKH to the liver and to hepatocellular carcinoma cells by being hydrolyzed without a reductive activation process and exhibit strong antitumor effects compared with Vitamin K [18–20]. Accordingly, these VKH derivatives may act as delivery systems for VKH to avoid photodegradation and phototoxicity. Also, Vitamin K (quinone-type) is the difficulty in dose adjustment owing to its high lipid solubility and insolubility in aqueous media. Since VKH derivatives are designed in the form of powders, they are comparatively easy to prepare. Moreover, VKH derivatives are hydrophilic and can be dispersed in aqueous media. These aqueous formulations can be used in in vitro experiments without using solubilizing agents such as surfactants.

Here, we assessed their photostability and phototoxicity in order to further development of a vitamin K skin application.

3.2 Evaluation of the photostability of VKH derivatives

The ethanol solutions of PK, MK-4, phyllohydroquinone derivatives (PKH-DMG and PKH-SUC), and menahydroquinone-4 derivatives (MKH-DMG and MKH-SUC) in quartz cells were exposed to artificial sunlight (12000 lx) from the vertical direction at 25°C with and without shading. The residual concentrations were determined by liquid chromatography tandem mass spectrometry [12, 13].

All samples were photodegraded according to the apparent first-order rate equation by artificial sunlight irradiation, and their apparent first order rate constants (*k*) and half-lives ($t_{1/2}$) of degradation are shown in **Table 1**. The concentrations of vitamin K (quinone form) and DMG ester derivatives were unchanged under shading, whereas the concentrations of SUC ester derivatives decreased both with and without shading. The decreased concentration of with shading was related to hydrolysis of the bis-ester to monoesters. The difference in the degree of hydrolysis for DMG and SUC ester derivatives are probably due to the stability of the ester bonds.

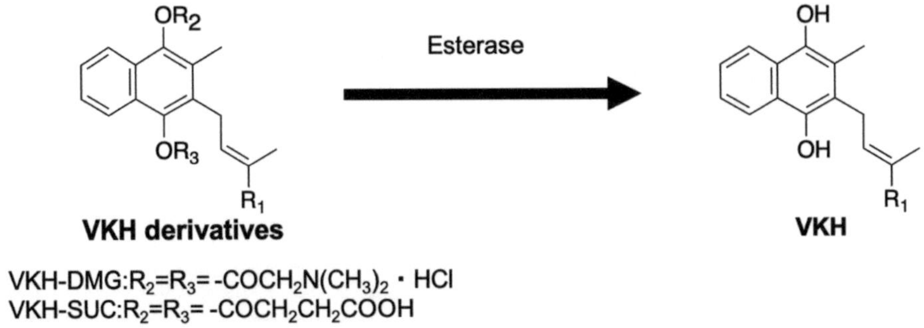

Figure 3.
Structure of the VKH derivatives and hydrolytic of VKH derivatives to VKH.

Compound[a]	Irradiation conditions	k (h^{-1})	$t_{1/2}$ (h)
PK	Sunlight	5.532	0.125
	Shading[b]	-[c]	-[c]
PKH-DMG	Sunlight	0.140	4.950
	Shading[b]	-[c]	-[c]
PKH-SUC	Sunlight	1.219	0.569
	Shading[b]	0.577	1.201
MK-4	Sunlight	8.239	0.084
	Shading[b]	-[c]	-[c]
MKH-DMG	Sunlight	0.167	4.150
	Shading[b]	-[c]	-[c]
MKH-SUC	Sunlight	2.796	0.248
	Shading[b]	0.883	0.785

[a]*The initial concentration was 1 μM in ethanol.*
[b]*During irradiation, the compound was covered with aluminum foil to provide shade.*
[c]*No decomposition.*

Table 1.
Apparent first order rate constants (k) and half-lives ($t_{1/2}$) of degradation of vitamin K and VKH derivatives in ethanol under irradiation using artificial sunlight (12000 lx) at 25°C.

The half-lives of PK and MK-4 irradiated with artificial sunlight were 0.125 and 0.08 h, respectively. Moreover, the half-lives of PKH-SUC and MKH-SUC were approximately 5- and 3-fold more stable than those of PK and MK-4, respectively, although the stability was not greatly improved. In contrast, the half-life of PKH-DMG was approximately 40-fold greater than that of PK, and the half-life of MKH-DMG was approximately 50-fold greater than that of MK-4, supporting that high light stability could be ensured against artificial sunlight. Note that no formation of chromenol was observed from irradiating the VKH derivatives.

The wavelength distribution of sunlight is wide from ultraviolet to infrared. To examine the wavelength characteristics of photodegradation, the photostability of quinone-type vitamin K and VKH derivatives was evaluated after irradiation with monochromatic light at 279, 341, 373, 404, or 435 nm. **Table 2** shows the photodegradation rate (k) and the irradiation energy of each wavelength at which the residual concentration reaches half ($E_{1/2}$). Photodegradation of PK and MK-4 occurred at all measured wavelengths (279–435 nm), and the decomposition rate accelerated with shorter wavelengths. By contrast, the photodegradation of PKH-DMG and MKH-DMG accelerated at a wavelength of 279 nm. In addition, the degradation of PKH-SUC and MKH-SUC occurred at all wavelengths, and the photodegradation rates were almost the same at wavelengths above 341 nm. Therefore, these findings clarified that decomposition at wavelengths above 341 nm involved hydrolysis to the monoester, but not photodegradation, and photodegradation of PKH-SUC and MKH-SUC was accelerated at a wavelength of 279 nm.

The above results clearly confirmed that the VKH derivative is more stable to sunlight than vitamin K (quinone form) and has a narrow wavelength range for photodegradation.

Compound[a]	Wavelength (nm)	k (J^{-1} × cm^2)	$E_{1/2}$ (J × cm^{-2})
PK	279	0.549	1.262
	341	0.359	1.933
	373	0.094	7.390
	404	0.026	26.260
	435	0.021	32.459
PKH-DMG	279	0.146	4.750
	341	-[b]	-[b]
	373	-[b]	-[b]
	404	-[b]	-[b]
	435	-[b]	-[b]
PKH-SUC	279	0.137	5.047
	341	0.070	9.889
	373	0.076	9.169
	404	0.078	8.860
	435	0.079	8.828
MK-4	279	0.533	1.301
	341	0.422	1.643
	373	0.151	4.583
	404	0.049	15.800
	435	0.035	19.738
MKH-DMG	279	0.146	4.750
	341	-[b]	-[b]
	373	-[b]	-[b]
	404	-[b]	-[b]
	435	-[b]	-[b]
MKH-SUC	279	0.110	6.323
	341	0.069	10.036
	373	0.059	11.792
	404	0.061	11.296
	435	0.068	10.253

[a]*The initial concentration was 1 μM in ethanol.*
[b]*No decomposition.*

Table 2.
The rate constants (k) and half-lives ($E_{1/2}$) of degradation of vitamin K and VKH derivatives in ethanol under different irradiation intensities of monochromatic light at 25°C.

3.3 Evaluation of phototoxicity

To confirm whether the VKH derivatives had phototoxic properties, singlet oxygen generation, intracellular ROS generation, and cytotoxicity after irradiation with UVA were evaluated in HaCaT cells [12, 13].

Figure 4 shows the amounts of singlet oxygen produced by each compound (200 μM) irradiated with UVA (15 J/cm^2) in phosphate-buffered saline (PBS).

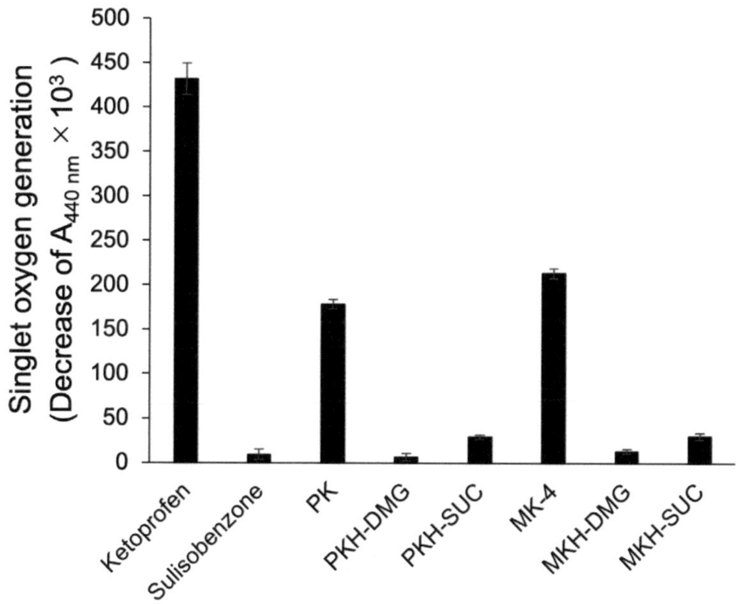

Figure 4.
Singlet oxygen generation in aqueous solutions of various compounds (200 μM) exposed to UVA (15 J/cm²). Data represent means ± standard deviations (n = 3).

Ketoprofen was used as a positive control, and sulisobenzone was used as a negative control. Vitamin K (quinone form) showed singlet oxygen generation depending on UVA irradiation energy, whereas VKH derivatives showed almost no singlet oxygen generation.

Analysis of intracellular ROS generation and cell viability following UVA irradiation at the time of MK-4, MKH-DMG, or MKH-SUC addition (50 μM) in HaCaT cells is shown in **Table 3**. MK-4 irradiated with UVA (5 J/cm²) increased intracellular ROS generation and decreased cell viability, whereas the MKH derivative did not. Similar trends were observed with PK and PKH derivatives [12].

Compound[a]	UVA irradiation intensity (J/cm²)	ROS generation (%control without UVA)	Cell viability[b] (%control without UVA)
Control	0	100 ± 6.19	100 ± 6.04
	5	275 ± 22.0	86.6 ± 5.44
MK-4	0	99.2 ± 6.87	104 ± 2.40
	5	1150 ± 131	11.3 ± 2.17
MKH-DMG	0	126 ± 6.63	104 ± 1.93
	5	282 ± 4.40	101 ± 2.00
MKH-SUC	0	125 ± 4.54	93.9 ± 4.79
	5	271 ± 15.5	88.8 ± 6.07

[a]*Cells were treated with PBS containing 50 μM MK-4, MKH-DMG, or MKH-SUC. Data represent means ± standard deviations (n = 3).*
[b]*Cell viability was measured at 24 h after UVA irradiation.*

Table 3.
Percent of intracellular ROS generation and cell viability in HaCaT cells in aqueous solutions of vitamin K and VKH derivatives with or without UVA irradiation (5 J/cm²).

As mentioned above, the photodegradation and phototoxicity of vitamin K (quinone form) is charge transfer from the β, γ-double bond of the isoprenyl side chain to the quinone moiety initiates intramolecular proton transfer from the side chain. Since VKH derivatives in which the two hydroxyl groups were protected by ester bond, it is considered that the charge transfer from the isoprenyl side chain that triggers a photochemical reaction was suppressed. These results strongly supported that VKH derivatives without a quinone structure did not show the same photodegradation and phototoxicity as vitamin K (quinone form) and may therefore be applied topically to the skin.

4. VKH delivery into HaCaT cells with vitamin K and VKH derivatives

To confirm whether VKH derivatives function as VKH prodrugs in skin-derived cells, the delivery properties of VKH to HaCaT cells were evaluated [12, 13]. **Table 4** shows the area under the curve (AUC) of intracellular VKO up to 72 h after the administration of MK-4 and VKH derivatives to HaCaT cells. VKO was used as an index of VKH because it is stoichiometrically produced from VKH after functioning as a cofactor for GGCX.

The $AUC_{VKO\ (0\text{-}72h)}$ values of PKH-DMG and PKH-SUC were 0.741- and 22.9-fold higher than that of PK, respectively. Additionally, the $AUC_{VKO\ (0\text{-}72h)}$ values of MKH-DMG and MKH-SUC were 1.02- and 1.64-fold higher than that of MK-4, respectively.

Based on these findings, vitamin K (quinone form) and VKH derivatives are converted to VKH in HaCaT cells and function as cofactors for GGCX. Thus, VKH derivatives can function as prodrugs of VKH. Furthermore, VKH derivatives could deliver VKH at concentrations equal to or higher than vitamin K (quinone form).

Compound[a]	**$AUC_{VKO(0\text{–}72h)}$ (nmol × h/mg of protein)**
PK	1.176 ± 0.056
PKH-DMG	0.872 ± 0.138
PKH-SUC	26.967 ± 2.030
MK-4	10.543 ± 0.795
MKH-DMG	10.786 ± 1.696
MKH-SUC	17.304 ± 1.068

[a]*Cells were treated with medium containing 5 μM compounds. Data represent means ± standard deviations (n = 3).*

Table 4.
Area under the curve over 72 h of VKO treated with PK, PKH derivatives, MK-4 or MKH derivatives in HaCaT cells.

5. Conclusion

Although many studies have supported the application of vitamin K for the treatment of skin pathologies, this compound is difficult to use as an external preparation to the skin owing to its photo-instability and phototoxic properties. The photodegradation and phototoxicity of vitamin K are derived from its quinone structure. Avoiding chromenol formation may suppress photodegradation via singlet oxygen and radical formation. Moreover, VKH derivatives in which the quinone structure is protected by ester bonds do not show chromenol formation and can be used to overcome the photo-instability and phototoxicity associated with vitamin

K while promoting VKH delivery to skin cells. Thus, VKH derivatives may be used for application to sites where shading may be difficult, as alternatives to vitamin K (quinone form).

Acknowledgements

This work was funded by JSPS KAKENHI (grant number: 20 K16066).

Conflict of interest

The authors declare no conflict of interest.

Author details

Shotaro Goto*, Shuichi Setoguchi, Kazuhisa Matsunaga and Jiro Takata
Faculty of Pharmaceutical Sciences, Fukuoka University, Fukuoka, Japan

*Address all correspondence to: sgoto@fukuoka-u.ac.jp

References

[1] Lou WW, Quintana AT, Geronemus RG, Grossman MC. Effects of topical vitamin K and retinol on laser-induced purpura on nonlesional skin. Dermatologic Surgery 1999;25: 942-944

[2] Lopes LB, Speretta FFF, Vitoria M, Bentley LB. Enhancement of skin penetration of vitamin K using monoolein-based liquid crystalline systems. European Journal of Pharmaceutical Sciences 2007;32:209-215

[3] Cohen JL, Bhatia AC. The role of topical vitamin K oxide gel in the resolution of postprocedural purpura. Journal of Drugs in Dermatology 2009;8:1020-1024

[4] Ocvirk J. Management of cetuximab-induced skin toxicity with the prophylactic use of topical vitamin K1 cream. Radiology and Oncology 2010; 44:265-266

[5] Pinta F, Ponzetti A, Spadi R, Fanchini L, Zanini M, Mecca C, et al. Pilot clinical trial on the efficacy of prophylactic use of vitamin K1-based cream (Vigorskin) to prevent cetuximab-induced skin rash in patients with metastatic colorectal cancer. Clinical Colorectal Cancer 2014;13:62-67

[6] Gaiser MR, Lorenzen S, Merx K, Trojan J, Ocvirk J, Ettrich TJ, et al. Evaluation of EGFR inhibitor-mediated acneiform skin toxicity within the double-blind randomized EVITA trial: A thorough gender-specific analysis using the WoMo score. Cancer Med 2019

[7] Hemmati AA, Houshmand G, Ghorbanzadeh B, Nemati M, Behmanesh MA. Topical vitamin K-1 promotes repair of full thickness wound in rat. Indian Journal of Pharmacology 2014;46:409-412

[8] Teraoka R, Matsuda Y. Stabilization-oriented preformulation study of photolabile menatetrenone (vitamin K2). International Journal of Pharmaceutics 1993;93:85-90

[9] Hangarter MA, Hormann A, Kamdzhilov Y, Wirz J. Primary photoreactions of phylloquinone (vitamin K-1) and plastoquinone-1 in solution. Photochemical and Photobiological Sciences 2003;2:524-535

[10] Official Journal of the European Union., 2009. Commission Directive 2009/6/EC of 4 February 2009 amending Council Directive 76/768/EEC, concerning cosmetic products, for the purpose of adapting Annexes II and III thereto to technical progress. https://eur-lex.europa.eu/legal-content/EN/TXT/HTML/?uri=CELEX:32009L0006&from=GA. (Accessed on 22 February 2021).

[11] Scientific Committees on Consumer Safety of the European Comission,. 2010. Opinion on Vitamin K1 (Phytonadione). https://ec.europa.eu/health/sites/health/files/scientific_committees/consumer_safety/docs/sccs_o_014.pdf. (Accessed on 22 February 2021).

[12] Goto S, Setoguchi S, Nagata-Akaho N, Terada K, Watase D, Yamakawa H, et al. Ester derivatives of phyllohydroquinone effectively deliver the active form of vitamin K-1 topically, owing to their non-photosensitivity. European Journal of Pharmaceutical Sciences 2020;155:8

[13] Goto S, Setoguchi S, Yamakawa H, Watase D, Terada K, Matsunaga K, et al. Prodrugs for skin delivery of menahydroquinone-4, an active form of vitamin K-2(20), could overcome the photoinstability and phototoxicity of vitamin K-2(20). International Journal of Molecular Sciences 2019;20:13

[14] Foote CS. DEFINITION OF TYPE-I AND TYPE-II PHOTOSENSITIZED

OXIDATION. Photochemistry and Photobiology 1991;54:659-

[15] Baptista MS, Cadet J, Di Mascio P, Ghogare AA, Greer A, Hamblin MR, et al. Type I and type II photosensitized oxidation reactions: Guidelines and mechanistic pathways. Photochemistry and Photobiology 2017;93:912-919

[16] Girotti AW, Kriska T. Role of lipid hydroperoxides in photo-oxidative stress signaling. Antioxidants and Redox Signaling 2004;6:301-310

[17] Ryu A, Arakane K, Koide C, Arai H, Nagano T. Squalene as a target molecule in skin hyperpigmentation caused by singlet oxygen. Biological and Pharmaceutical Bulletin 2009;32: 1504-1509

[18] Takata J, Karube Y, Hanada M, Matsunaga K, Matsushima Y, Sendo T, et al. VITAMIN-K PRODRUGS. 1. Synthesis of amino acid esters of menahydroquinone-4 and enzymatic reconversion to an active form. Pharmaceutical Research 1995;12:18-23

[19] Takata J, Karube Y, Hanada M, Matsunaga K, Matsushima Y, Sendo T, et al. Vitamin K prodrugs. 2. Water-soluble prodrugs of menahydroquinone-4 for systemic site-specific delivery. Pharmaceutical Research 1995;12: 1973-1979

[20] Takata J, Karube Y, Hanada M, Matsunaga K, Iwasaki H. Prodrug for bioreductive activation-independent delivery of menahydroquinone-4: Human liver enzymatic activation and its action in warfarin-poisoned human liver. Biological and Pharmaceutical Bulletin 1999;22:172-178

Chapter 6

Vitamin K Deficiency and Vascular Calcification. Is There Any Evidence about Its Impact on Coronary Artery Disease?

Theodoros Michailidis, Asterios Karakanas, Nikolaos Schizas and Petros Keryttopoulos

Abstract

Nowadays cardiovascular disease remain globally the leading cause of mortality. Coronary artery disease is the predominant clinical entity related to fatal cardiovascular events, while its development is mostly associated with progressive atherosclerosis of the vessels combined with gradual vascular calcification. It is well described and understood that vascular calcification is strongly associated with the occurrence of CVD and increased mortality rates. Therefore, it is essential to understand the metabolic pathways leading to its formation in order to develop effective therapies. A group of vitamin-k dependent proteins seems to play a significant role on the prevention of the arterial wall. Several past studies have shown that in cases of vitamin-k deficiency the process of vessel calcification is accelerated. Vitamin-k depletion and high levels of uncarboxylated and dephosphorylated forms of the aforementioned proteins are considered as important factors that contribute significantly to this rapid progression. Promising studies are giving the stimulus for further research in the field of vitamin-k supplementation and the suspension of vascular calcification.

Keywords: vitamin K, deficiency, vitamin-K dependent proteins, vascular calcification, coronary artery disease

1. Introduction

Cardiovascular diseases are the leading cause of mortality worldwide. Their constant increasing rate is attributed to many factors as the adoption of poor dietary habits and sedentary lifestyle accompanied in some cases with hereditary background. The so-called Western way of living has led also to the rise of other diseases such as dyslipidemia, hypertension, obesity, diabetes mellitus and chronic kidney disease, which are the main risk factors of CVD. Those conditions are mainly responsible for the rapid progression of vessel's atherosclerosis and plaque formation. Intimal calcification of large vessels is the main effect of those disorders responsible for plaque's progression. However, atherosclerosis is not the only factor. Vascular calcification is another major and independent risk factor strongly related with the development of vessel plaque leading to CVD [1]. It is a chronic

IntechOpen

and multi-factorial procedure related mainly with disorders such as chronic kidney disease and diabetes mellitus [2]. The metabolic pathways which are responsible for the progression of vascular calcification is not well-established and consequently treatment remains a challenge. Media arterial layer is the target area of calcification and can involve all vessel sizes regardless of the concurrence of atheromatic plaque [3]. The negative effect of this procedure can gradually lead to the development of valvural heart disease and coronary artery disease (CAD).

Matrix degradation and modification is the main reason of medial arterial calcification. Matrix Gla protein (MGP), Gla rich protein (GRP) and growth arrest specific gene-6 (Gas-6) are a group of vitamin-k dependent proteins that is considered to effectively inhibit the progression of vascular calcification. For example, in cases of vitamin-k deficiency, glutamic acid residues do not convert to the amino acid γ-carboxyglutamic acid residuals (Gla) which is a vital part for the normal function of MGP. Consequently, inactive forms of MGP (dp-ucMGP) accumulate in calcified tissues such as the blood vessels. Past studies have proposed ucMGP as a potential marker of vitamin-K deficiency, vascular calcification and cardiovascular disease [4]. Other trials have showed that patients under treatment with vitamin-k antagonists (VKAs) had higher scores of vascular calcification compared to a group of controls. Animal trial connected the administration of VKAs with an accelerated progression of medial calcification. On the other hand, when high doses of vitamin-K were administered, calcification lesions started slowly to improve [5]. Perspectives about the benefit of vitamin-K administration in specific population remain controversial, and therefore further research is required in this domain.

2. Vitamin-K dependent proteins

The correlation between low vitamin-K levels and rapid progression of vascular mineralization is supported by researchers in the past [6]. The pathophysiological pathways that justify such claims involve a group of vitamin-K dependent protein and calciprotein particles. Those proteins are the matrix Gla protein (MGP), gamma-carboxylated Gla-rich protein (GRP) which are both considered as strong inhibitors of vascular calcification and the growth specific arrest gene-6. Matrix Gla protein (MGP) is a vitamin-k2 dependent protein which is secreted by many cells, including the vascular smooth muscle cells (VSMC) and has high affinity to calcium crystals. Its effect on atherosclerotic plaque may probably derive from the blocking of calcium accumulation. It's present in bones and cartilage, in kidneys but also in blood vessels, endothelium and coronary artery. Matrix Gla protein prevents from the calcification development in soft tissues such as the blood vessels. Additionally, its preventive effect is exerted by inhibiting the bone morphogenetic protein-2 and 4 (BMP-2, BMP-4). This effect is of outmost importance because it prevents the VSMC from the transdifferentiation into osteoblasts like-cells which lead into progressive calcification of atherosclerotic plaques. In order to exert its effect, MGP has to be synthesized locally in the VSMCs. Vitamin-k2 is used as co-factor to its formation through the post-translational γ-carboxylation and phosphorylation of inactive MGP [7]. Carboxylated MGP (cMGP) is the active protein form responsible for the prevention of the calcification of the arterial wall [8]. MGP's expression from the VSMCs and the endothelium may give us the opportunity to measure its circulating levels. As it was mentioned before, cMGP has high affinity to calcium crystals and binds strongly to them preventing from their accumulation into the arterial wall. Considering that cMGP could evolve into a very promising biomarker for the prognosis and the progression of calcification as well as a prognostic factor to severe cardiovascular outcomes [9]. On the other hand, it should be mentioned that inactive forms of MGP have been associated with accelerated rate of vascular calcification but also with increased mortality [10]. Whereas the active form of MGP is both carboxylated and phosphorylated, the inactive forms have not undergone one of those two metabolic steps or both of them (uncarboxylated, dephosporylated MGP, dp-ucMGP). cMGP is the only factor that under specific conditions may promote the reversal of vascular calcification [11]. It is worth mentioning that calcified arteries have high concentrations of MGP and the severity of such lesions was related to MGP serum levels [12]. Gla rich proteins is another vitamin K dependent protein which is present in both bone and cartilage and also has high affinity to calcium. High concentrations of γ-carboxylated GRP have been observed in individuals with increased vascular calcification. However, suggested data derive mainly from animal studies and its functionality in humans is not well established. Calciprotein particles are particles that prevent from vascular calcification by blocking the formation of the calcium/phosphate crystals. These particles also contain high levels of MGP and GRP, so in cases of vitamin-k deficiency, their effect may be impaired. Vascular smooth muscle cells excrete also the growth arrest specific gene 6 protein (Gas-6) which is one of the stimulating factors of their growth. It exerts its preventive effect through several pathways. Initially it stimulates the bcl-2 anti-apoptotic protein which is responsible for the inhibition of caspase-3. Caspase-3 is a crucial protein for the pro-apoptotic cell procedure, so through that way Gas-6 protects VSMCs from transdifferentation into cells with osteoblast-like effects. Apoptotic cells could be used as substrate for the development of constant inflammation and excessive calcification. Thus, Gas-6 through this way Gas-6 prevents the arterial from this process. However, in order to exert

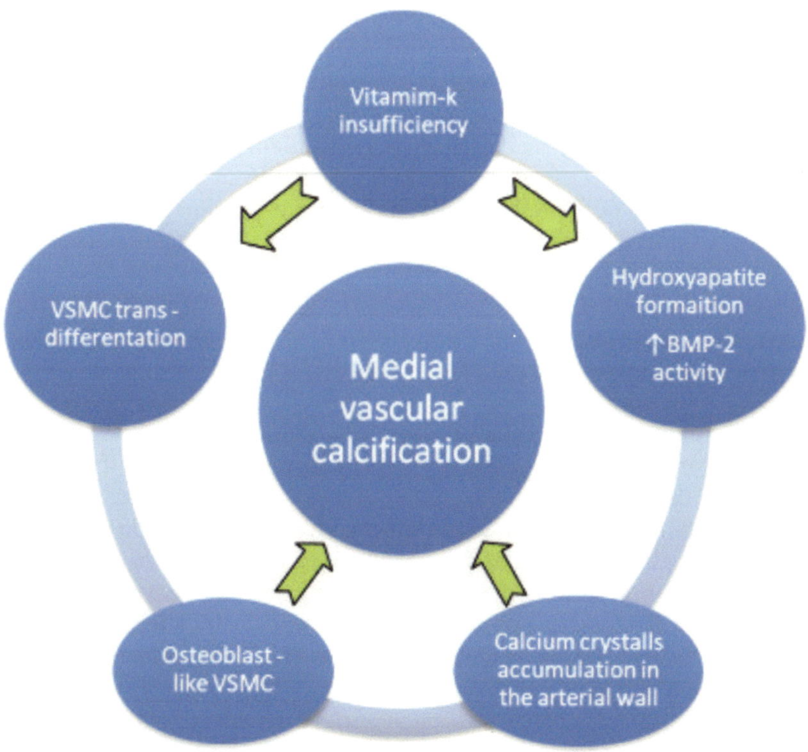

Figure 1.
Summary of the effect of vitamin K deficiency on vascular calcification. Abbreviations: BMP-2, bone morphogenetic protein-2; VSMCs, vascular smooth muscle cell.

its effects Gas-6 needs γ-carboxylation a process in which vitamin-k is a necessary co-factor [13]. Taking all that into consideration, Gas-6 is vitamin-K dependent and very important factor, against the VMSCs apoptosis and the progression of calcified plague. In cases of vitamin-k deficiency the lack of Gla proteins affect negatively the growth of VSMC [14] (**Figure 1**).

3. Postmenopausal women, vitamin-k and vascular calcification

In past studies it was also observed that post-menopausal women with osteoporosis had increased rates of arterial calcification [15]. Based on that, researchers have investigated the potential correlation of vitamin-k status, bone formation rate and vascular calcification of post-menopausal women. Initially, lower bone formation rate in parallel correlation with low levels of serum vitamin-k. The same also applies for the increased frequency of vascular calcification observed in post-menopausal women [16]. Concerning, the lower bone mass of post-menopausal women, it seems that vitamin-k levels and osteocalcin induced remodeling bone process are closely related. Though not with the same metabolic way to vascular calcification but in both situations vitamin-k is necessary [17]. It is important to mention that few trials, evaluated the potential benefit of vitamin k2 supplementation in post-menopausal women with aortic calcification and osteoporotic lesions. According to the authors, calcified lesions were at least sustained or in some cases decreased. However, these beneficial effects were observed in individuals that received vitamin-k2 supplementation but not vitamin-k1 [18]. Finally, results from the study of Gast et al. in this group of patients, have showed that vitmamin-k2 had the greatest preventive effect in CAD [19]. There are data that suggest that osteoporotic women are on higher risk of cardiovascular disease compared to match-aged non-osteoporotic individuals. These evidence imply that this difference occurs due to increased vascular calcification of both large vessels and coronary artery in those patients. Vitamin-k2 may have an essential role both to bone formation via its effect on osteocalcin as well as to the reduction of vascular mineralization. Further and larger studies are needed in order to assess the potential benefit of vitamin-k2 supplementation in both bone mass loss, atherosclerotic progression and prevention of cardiovascular disease in this specific population [20].

4. Vitamin-k-antagonists and vascular calcification

Vitamin-k antagonists are oral anticoagualants that reduce the levels of vitamin-k by interfering in vitamin's recycle. They essentially inhibit vitamin-k reductase complex-1 and consequently reducing active vitamin-k reserves. Thus, they prevent from the formation of blood clots. Nevertheless, vitamin-k depletion in serum has some detrimental effect on many sequences of steps in metabolic pathways like those mentioned before. Taking that in mind, previous studies have demonstrated the negative effect of VKAs on the progress of vascular calcification [5]. Win et al. examined the potential benefit of apixaban administration in patients with atrial fibrillation in terms of the atherosclerotic plaque progression. Study results showed that apixaban administration was associated with slower plaque progression as well as lower calcium scores compared to warfarin treated group. Taking into consideration the usual co-existence of atrial fibrillation and coronary artery disease, apixaban could be a preferable choice in patients with atrial fibrilation not only for the prevention of thromboembolic events but also to slow down coronary plague progression and prevent from fatal cardiovascular events [21]. Accordingly, outcomes from a trial that compared the use rivaroxaban against warfarin have also indicated the beneficial effect of NOAGs into the slower plaque progression arterial calcification [22]. Therefore, in cases of individuals with established CAD or high level of vascular atherosclerosis in need for anticoagulation, the physician should choose wisely the appropriate treatment. These data suggest the preferable choice of NOACs instead of VKAs though further studies are needed for more robust results. However, what about patients that are in need for VKAs coagulation, or the use of NOACs is contraindicated? Such dilemmas are common in clinical practice and therefore the presence of clearly defined guidelines is essential.

5. Chronic kidney disease, vitamin-k status and vascular calcification

Chronic kidney disease is strongly associated with the presence of increased cardiovascular risk. The most frequent cause of death in this population is coronary artery disease as well as stroke events. Hypertension, dyslipidemia, diabetes mellitus, which are common comorbidities in patients with renal failure, are important factors that can lead to excessive atherosclerosis and the development of CVD. Moreover, vascular calcification is also excessive in patients with chronic kidney disease (CKD) and so it constitutes an independent risk factor for the development of CVD [23]. Age, time on hemodialysis, persistent hyperphospahetamia and hypercalciemia are some of the causes of accelerated vascular calcification. High calcium and phosphate levels are associated with the deposition of hydroxyapatite into the arterial wall [24]. Whereas the presence of intimal calcification, which is related mainly to large vessels, is almost the same between individuals with or without end-stage renal disease, this is not the case for median calcification of the arterial wall. For example, aorta calcification is worse in patients with renal disease compared to those without. Aorta calcification increases arterial stiffness, which consequently aggravates the present hypertension and finally contributes to the development or the deterioration of a pre-existing left ventricular hypertrophy and left ventricular insufficiency. Patients with end-stage renal disease have increased vascular calcification compared to non-CKD individuals, especially those that are under hemodialysis [25]. Coronary artery calcification is common among older patients with CKD. However past studies have also showed that younger individuals with end-stage renal disease have also a high percentage of vascular calcification as well as coronary artery calcification [26]. Using high resolution computed tomography; researchers have proved that coronary calcification was higher in young adults that were on dialysis compared to those that were not. At this moment, the main treatment approach against the vessel mineralization of CKD patients is the strict regulation of calcium and phosphate balance. Additionally, clinician focus their attention to better treatment management of concomitant disorders that affect atherosclerotic plaque, such as diabetes, dyslipidemia and smoking cessation. Even though there is no strong evidence that support the administration of agents that inhibit calcification, it is easy to understand that those interventions might play a crucial role to the prevention of CVD in CKD-individuals. In a microenvironment of constant inflammation, vascular smooth cells are gradually transformed in osteoblasts like-cells promoting then the development of medial artery calcification. This process is mediated by multiple proteins and is facilitated by the presence of systemic or local inflammation. The latter is of outmost importance for the reason that macrophages excrete among others, matrix-metalloproteinase which lead to the apoptosis of elastic fibers and thus promote vascular calcification. In addition, vitamin k-dependent matrix Gla protein and Gia-rich protein are both important for vascular calcification. As it was mentioned before, low levels of those proteins have been associated with increased rate of calcification and development of cardiovascular disease. Several studies in the past have proved the increased coronary artery calcification in CKD-individuals by calculating the number of vessel calcifications using high resolution computed tomography. According to the study of Holden et al. in patients with CKD and/or end-stage renal disease vitamin-K deficiency is very common. It is possible that this outcome has some extra effect on the progressive calcification of these patients [27]. At this point, treatment in order to delay the progression of vascular calcification in ESRD and CKD patients targets to the regulation of calcium and phosphate. However, dietary advice in order to prevent hyperkaliemia or hyperphospatemia leads to the avoidance of food rich in vitamin k, inducing the existing deficiency. A recent study with hemodialysis patients have studied the ffect of vitamin-k supplementation on the plasma levels of ucMGP, uncarboxylated osteocalcin and PIVKA-II. The results were very promising, because have showed a significant decrease of the inactive form of MGP and also proved the vitamin-k deficiency in this specific population. The outcomes of this study may be the beginning of future randomized controlled trials that will evaluate the effect of vitamin-k supplementation on vascular calcification of CKD patients [28]. Respectively, another study by Oikonomaki et al. in hemodialysis patients have demonstrated the reduction in uc-MGP levels after 1-year of vitamin-k2 supplementation but the progression of vascular calcification remained the same between studied groups [29]. It is possible that only vitamin-k supplementation is not enough in order to overt calcification. This process in such individuals is so multi-factorial that need a comprehensive treatment approach that will slow the progression. Further studies are needed in order to clarify the benefits of vitamin-k supplementation in ESRD or CKD patients.

6. Vitamin-k and coronary calcification

Vascular calcification and especially coronary calcification is a strong predictor of coronary events and this is a process that is regulated actively by vitamin K dependent proteins which are called matrix Gla proteins. There are currently no pharmacological means to improve vascular stiffness and vascular calcification. There is growing evidence that vitamin K, a cheap and safe intervention that can have beneficial effects on cardiovascular health. Vitamin K straightforward administration can reduce the progression of vascular calcification. The biological rationale is that supplementation with vitamin K will carboxylate (activate) Gla proteins, whose role, among others, is to reduce the progression of vascular calcification [30]. Coronary artery calcification which is a significant marker of cardiovascular disease is affected by these dependent vitamin K proteins. As it was mentioned before vitamin-K deficiency promotes coronary artery calcification [6]. In animal studies, MGP removal showed severe progression of vessel calcification and this empowered the theory about vitamin K role in this process [31]. The most of the clinical trials showed that supplementation with Vitamin K2 (menaquinone) had beneficial effects in cardiovascular calcification. But a clinical trial from Shea et al. showed that also supplementation with vitamin K1 (phyloquinone) slowed the progression of coronary artery calcification. Though its effect was enhanced by the parallel administration of vitamin-D and calcium supplements [32]. On the other hand, Beulens et al., with a cross-sectional study among 564 post-menopausal women reached the conclusion that only high intake of menaquinone is associated with reduced coronary calcification, as it was measured via computed cardiac tomography [31]. Also Vossen et al. wanted to study a sample of patients with coronary artery disease and follow them up via Agatston calcium score about the progression of coronary calcification as long as the individuals were under vitamin-k supplementation [33]. The Roterdam Study, a prospective population-based study, showed in a sample of 7983 men and women aged 55 y and over in a follow up to 10 years, that dietary intake of menaquinone had a protective effect against coronary heart disease [18]. Also a meta analysis of 3 US Cohorts, among 3891 patients, who measured fasting circulating phyloquinone levels, showed that low phyloquinone levels was associated with increased all – cause mortality but not of CVD [32]. Many clinical trials as referred above showed a possible correlation between vitamin K and increased artery calcification mainly in high risk patients. A systematic review from Hartley et al. tried to show if there is primary prevention from CVD in healthy individuals who received supplementation with vitamin K. They only found a small clinical trial with only 60 patients that fulfilled their criteria and there was no significant impact in primary prevention of CVD and other CVD factors such as blood pressure and plasma lipids level (Hartley). However, this was a small clinical trial with short duration. Further studies are necessary about the benefit of vitamin-k supplementation in the primary prevention of CVD.

7. Vitamin-K and valvular calcification

Valvular calcification is a common degenerative disease characterized by progressive valvurar calcification. Nowadays the incidence of valvural disease is higher probably due to increased life expectancy. The most commonly calcified valves is aortic valve followed by mitral valve. Based on the role of vitamin K as protective agent for cardiovascular health and especially as a protector against vascular calcification, some trials evaluated the potential effect of vitamin-k as a protective factor against the development of valvular calcification. Bradenburg et al. with a small prospective, open label clinical trial with 99 patients, selected individuals with asymptomatic or mild symptoms with aortic calcified valve and separated them in two study arms with vitamin-k1 supplementation and placebo. Then, they calculated the amount of calcification of the valve via cardiac computed tomography, at the beginning of the trial and after 12 months. Also they measured the dephosphorylated undecarboxylated MGP as a circulating marker for vitamin-k deficiency. Over the 12 month period, aortic valve calcification volume score progressed 10% in the arm with vitamin K supplementation compared with 22% in the placebo group. Also plasma dp-ucMGP were significantly reduced by 45% in the vitamin K group. On the other hand, this trial had many limitations such as the small sample, the short duration of follow up, the open-label design and the broad spectrum of severity of valvural disease at baseline [34]. Another clinical trial that is planned and want to study the effect of vitamin K in aortic calcification is from Peeters et al. concerning especially the calcification progress of bicuspid aortic valve. Bicuspid aortic valve, a common congenital abnormality, occurring in 13,7 per 1000 people in general population is associated with early development of calcific aortic valve stenosis. In this double-blind study they will supply vitamin K2 and follow up 44 people in a period over 18 months. The follow up of the sample will be every 6 months with PET/cardiac MRI, cardiac CT and echocardiography. This trial will provide us with more information for calcium activity on aortic valve and the potential effect of vitamin K. It will also open the way for large scale randomized clinical trials in order to develop potential treatment option against the progression of calcific aortic valve stenosis [35].

8. Vitamin-k supplementation and vascular calcification

Vitamin-K deficiency is associated with low-levels of MGP and Gas-6 and the accumulation of GRP, ucMGP and dpMGP forms in soft tissues and vessels. As it was described before, vitamin-K is used as co-factor for the post-translational γ-carboxylation of MGP and Gas-6 in order to maintain preventive effect on vessels and VSMCs respectively. In addition, vitamin-k deficiency is associated with higher circulating levels of uncarboxylated or dephosphorylated forms of MGP. A three arm randomized controlled trial assessed the levels of deposhpo-carboxylated MGP levels after 12 weeks of menaquinone-7 supplementation. The dp-ucMGP levels were reduced significantly. It is important to mention that according to the authors the outcome was dose-dependent and increased in time [4]. In a systematic review meta-analysis of trials with different study-design by Lees et al., dp-ucMGP levels were significantly reduced with the administration of vitamin-k. However, that effect did not lead to the improvement of vascular calcification with the exception of a small number of studies. Concerning the use of dp-ucMGP as biomarker of CVD the results are controversial and so that is not suggested. It is important to mention that the greater efficacy was achieved with vitamin-k2 supplementation rather than k1 [30]. These data suggest that dp-ucMGP could be an important marker of vitamin-k status. It is worth mentioning that the co-administration of vitamin-k and vitamin-D have showed some promising results concerning the prevention of fatal CVD. However, larger randomized controlled trials are needed in order to delineate if vitamin-k supplementation alone or in combination with vitamin-D could benefit patients with progressive vascular calcification [36] (**Table 1**).

	Vitamin-K sufficiency	**Vitamin-K deficiency (ex. Inadequate intake, VKAs administration)**
MGP (vitamin-K dependent carboxylation)	↑ active cMGP prevents: 1. The hydroxyapatite formation 2. The accumulation of calcium crystals 3. Transdifferentiation of VSMCs into osteoblast like-cells 4. Decreased BMP-2 activity	Significantly lower levels of MGP-2 and accumulation of its inactive forms (dp-uc MGP) with devastating impact on VC
Gas-6 protein (γ-carboxylation with vitamin-K as co-factor)	1. Activation of anti-apoptotic protein Bcl-2 2. Inhibition of pro-apoptotic protein Caspase-3.*	↓Gas-6 levels, inhibition of VSMCs growth and increased apoptosis
VSMCs	Inhibition of apoptosis and transdifferention into osteoblast like-cells	Increased apoptosis and transdifferention into osteoblast like-cells
Vascular Calcification	Suspended and possibly reversed medial VC**	Increased medial VC

VSMCs status and progression of vascular calcification depending on vitamin-k serum levels. Abbreviations: MGP, matrix Gla protein; Gas-6 protein, growth arrest specific gene 6 protein; VSMCs, vascular smooth muscle cell; VC, vascular calcification.
**Apoptotic cells are used as substrate for accumulation of calcium crystals.*
***Observed in small trials with co-administration with vitamin D. Further studies are needed.*

Table 1.
Summary of changes in vitamin-k dependent proteins.

9. Conclusion

Vitamin-k is a fat-soluble vitamin mostly known for its significant role in the coagulation sequence of steps. However, vitamin-k is also important to other also important metabolic pathways. Vascular calcification has proved to be a multi-factorial procedure that leads to the transdifferentation of VSMCs into osteoblast phenotype like-cells and a vicious circle of arterial wall calcification. Vitamin-K dependent proteins MGP, GRP and Gas-6 play an important role in the regulation of this constant progressive process. In cases of vitamin-k deficiency, the preventive effect of those Gla proteins on the arterial wall declines and consequently the process of vascular calcification is enhanced. The calculation of the uncarboxylated and dephosphorylated forms of those proteins are considered a marker of vitamin-k deficiency. However, existing data do not suggest the use of dp-ucMGP as predicting factors of cardiovascular disease. Vitamin-k supplementation have been associated with a significant reduction in the dp-uc MGP circulating levels, although this reduction was not related with a reduction in the process of vascular calcification. Due to the multifactorial process for the formation of calcified plagues, it cannot be suggested at this point that vitamin-k supplementation alone will reverse those lesions. Emerging evidence support the co-administration of vitamin-D and vitamin-K has a greater effect against the progression of vascular calcification. Concerning the effect of vitamin-k deficiency and cardiovascular health, evidence remains controversial. It is a given that more studies are needed in this area in order to draw safe and robust conclusions. Vascular calcification and atherosclerotic plaques are strongly related with arterial stiffness, hypertension and impaired cardiac function. It is of outmost importance to find solution of this degenerative procedure in order to develop additional preventive and therapeutic strategies against the development of cardiovascular diseases.

Author details

Theodoros Michailidis[1]*, Asterios Karakanas[2], Nikolaos Schizas[3] and Petros Keryttopoulos[4]

1 Second Department of Internal Medicine, Ippokration General Hospital of Thessaloniki, Greece

2 Department of Cardiology, Papageorgiou General Hospital of Thessaloniki, Greece

3 Cardiovascular and Thoracic Department, Evangelismos General Hospital, Athens, Greece

4 Private Medical Practice, Veria, Greece

*Address all correspondence to: thgmichailidis@gmail.com

References

[1] R. J. M. W. Rennenberg, A. G. H. Kessels, L. J. Schurgers, J. M. A. Van Engelshoven, P. W. De Leeuw, and A. A. Kroon, "Vascular calcifications as a marker of increased cardiovascular risk: A meta-analysis," Vasc. Health Risk Manag., vol. 5, pp. 185-197, 2009, DOI:10.2147/vhrm.s4822.

[2] M. Wu, C. Rementer, and C. M. Giachelli, "Vascular calcification: An update on mechanisms and challenges in treatment," Calcif. Tissue Int., vol. 93, no. 4, pp. 365-373, 2013, DOI:10.1007/s00223-013-9712-z.

[3] L. L. Demer and Y. Tintut, "Vascular calcification: Pathobiology of a multifaceted disease," Circulation, vol. 117, no. 22, pp. 2938-2948, 2008, DOI:10.1161/CIRCULATIONAHA.107.743161.

[4] G. Schlieper et al., "Circulating nonphosphorylated carboxylated matrix Gla protein predicts survival in ESRD," J. Am. Soc. Nephrol., vol. 22, no. 2, pp. 387-395, 2011, DOI:10.1681/ASN.2010040339.

[5] B. Weijs et al., "Patients using vitamin K antagonists show increased levels of coronary calcification: An observational study in low-risk atrial fibrillation patients," Eur. Heart J., vol. 32, no. 20, pp. 2555-2562, 2011, DOI:10.1093/eurheartj/ehr226.

[6] M. Kyla Shea and R. M. Holden, "Vitamin K status and vascular calcification: Evidence from observational and clinical studies," Adv. Nutr., vol. 3, no. 2, pp. 158-165, 2012, DOI:10.3945/an.111.001644.

[7] C. M. Shea, C. M. Edgar, T. A. Einhorn, and L. C. Gerstenfeld, "BMP treatment of C3H10T1/2 Mesenchymal stem cells induces both Chondrogenesis and Osteogenesis," J. Cell. Biochem., vol. 90, no. 6, pp. 1112-1127, 2003, DOI:10.1002/jcb.10734.

[8] M. Murshed, T. Schinke, M. D. McKee, and G. Karsenty, "Extracellular matrix mineralization is regulated locally; different roles of two gla-containing proteins," J. Cell Biol., vol. 165, no. 5, pp. 625-630, 2004, DOI:10.1083/jcb.200402046.

[9] E. G. H. M. Van Den Heuvel et al., "Circulating uncarboxylated matrix Gla protein, a marker of vitamin K status, as a risk factor of cardiovascular disease," Maturitas, vol. 77, no. 2, pp. 137-141, 2014, DOI:10.1016/j.maturitas.2013.10.008.

[10] T. Ueland et al., "Undercarboxylated matrix Gla protein is associated with indices of heart failure and mortality in symptomatic aortic stenosis," J. Intern. Med., vol. 268, no. 5, pp. 483-492, 2010, DOI:10.1111/j.1365-2796.2010.02264.x.

[11] B. D. Parker et al., "The associations of fibroblast growth factor 23 and uncarboxylated matrix Gla protein with mortality in coronary artery disease: The heart and soul study," Ann. Intern. Med., vol. 152, no. 10, pp. 640-648, 2010, DOI:10.7326/0003-4819-152-10-201005180-00004.

[12] C. M. Shanahan, N. R. B. Cary, J. C. Metcalfe, and P. L. Weissberg, "High expression of genes for calcification-regulating proteins in human atherosclerotic plaques," J. Clin. Invest., vol. 93, no. 6, pp. 2393-2402, 1994, DOI:10.1172/JCI117246.

[13] D. Proudfoot, J. N. Skepper, L. Hegyi, M. R. Bennett, C. M. Shanahan, and P. L. Weissberg, "Apoptosis regulates human vascular calcification in vitro: Evidence for initiation of vascular calcification by apoptotic bodies," Circ. Res., vol. 87, no. 11, pp. 1055-1062, 2000, DOI:10.1161/01.RES.87.11.1055.

[14] B. K. Son et al., "Statins protect human aortic smooth muscle cells from

inorganic phosphate-induced calcification by restoring Gas6-Axl survival pathway," Circ. Res., vol. 98, no. 8, pp. 1024-1031, 2006, DOI:10.1161/01.RES.0000218859.90970.8d.

[15] C. E. Lampropoulos et al., "Osteoporosis and vascular calcification in postmenopausal women: A cross-sectional study," Climacteric, vol. 19, no. 3, pp. 303-307, 2016, DOI:10.3109/13697137.2016.1164134.

[16] K. S. G. Jie, M. L. Bots, C. Vermeer, J. C. M. Witteman, and D. E. Grobbee, "Vitamin K status and bone mass in women with and without aortic atherosclerosis: A population-based study," Calcif. Tissue Int., vol. 59, no. 5, pp. 352-356, 1996, DOI:10.1007/s002239900139.

[17] S. Urayama et al., "Effect of vitamin K2 on osteoblast apoptosis: Vitamin K2 inhibits apoptotic cell death of human osteoblasts induced by Fas, proteasome inhibitor, etoposide, and staurosporine," J. Lab. Clin. Med., vol. 136, no. 3, pp. 181-193, 2000, DOI:10.1067/mlc.2000.108754.

[18] J. M. Geleijnse et al., "Dietary intake of menaquinone is associated with a reduced risk of coronary heart disease: The Rotterdam study," J. Nutr., vol. 134, no. 11, pp. 3100-3105, 2004, DOI:10.1093/jn/134.11.3100.

[19] R. Caluwé, S. Vandecasteele, B. Van Vlem, C. Vermeer, and A. S. De Vriese, "Vitamin K2 supplementation in haemodialysis patients: A randomized dose-finding study," Nephrol. Dial. Transplant., vol. 29, no. 7, pp. 1385-1390, 2014, DOI:10.1093/ndt/gft464.

[20] L. A. J. L. M. Braam, A. P. G. Hoeks, F. Brouns, K. Halmuyák, M. J. W. Gerichhausen, and C. Vermeer, "Beneficial effects of vitamins D and K on the elastic properties of the vessel wall in postmenopausal women: A follow-up study," Thromb. Haemost., vol. 91, no. 2, pp. 373-380, 2004, DOI:10.1160/th03-07-0423.

[21] T. T. Win et al., "Apixaban versus warfarin in evaluation of progression of atherosclerotic and calcified plaques (prospective randomized trial)," Am. Heart J., vol. 212, pp. 129-133, 2019, DOI:10.1016/j.ahj.2019.02.014.

[22] J. Lee et al., "Randomized trial of rivaroxaban versus warfarin in the evaluation of progression of coronary atherosclerosis," Am. Heart J., vol. 206, pp. 127-130, 2018, DOI:10.1016/j.ahj.2018.08.007.

[23] R. B. Roijers, N. Debernardi, J. P. M. Cleutjens, L. J. Schurgers, P. H. A. Mutsaers, and G. J. Van Der Vusse, "Microcalcifications in early intimal lesions of atherosclerotic human coronary arteries," Am. J. Pathol., vol. 178, no. 6, pp. 2879-2887, 2011, DOI:10.1016/j.ajpath.2011.02.004.

[24] M. Fukagawa and J. J. Kazama, "The making of a bone in blood vessels: From the soft shell to the hard bone," Kidney Int., vol. 72, no. 5, pp. 533-534, 2007, DOI:10.1038/sj.ki.5002440.

[25] M. K. Sigrist, M. W. Taal, P. Bungay, and C. W. McIntyre, "Progressive vascular calcification over 2 years is associated with arterial stiffening and increased mortality in patients with stages 4 and 5 chronic kidney disease," Clin. J. Am. Soc. Nephrol., vol. 2, no. 6, pp. 1241-1248, 2007, DOI:10.2215/CJN.02190507.

[26] Goodman, "Coronary artery calcification in young adults undergoing dialysis," Young, vol. 342, no. 1478-1483, pp. 7-12, 2000.

[27] R. M. Holden, A. R. Morton, J. S. Garland, A. Pavlov, A. G. Day, and S. L. Booth, "Vitamins K and D status in stages 3-5 chronic kidney disease," Clin. J. Am. Soc. Nephrol., vol. 5, no. 4, pp.

590-597, 2010, DOI:10.2215/CJN.06420909.

[28] R. Westenfeld et al., "Effect of vitamin K 2 supplementation on functional vitamin K deficiency in hemodialysis patients: A randomized trial," Am. J. Kidney Dis., vol. 59, no. 2, pp. 186-195, 2012, DOI:10.1053/j.ajkd.2011.10.041.

[29] T. Oikonomaki et al., "The effect of vitamin K2 supplementation on vascular calcification in haemodialysis patients: A 1-year follow-up randomized trial," Int. Urol. Nephrol., vol. 51, no. 11, pp. 2037-2044, 2019, DOI:10.1007/s11255-019-02275-2.

[30] J. S. Lees, F. A. Chapman, M. D. Witham, A. G. Jardine, and P. B. Mark, "Vitamin K status, supplementation and vascular disease: A systematic review and meta-analysis," Heart, vol. 105, no. 12, pp. 938-945, 2019, DOI:10.1136/heartjnl-2018-313955.

[31] J. W. J. Beulens et al., "High dietary menaquinone intake is associated with reduced coronary calcification," Atherosclerosis, vol. 203, no. 2, pp. 489-493, 2009, DOI:10.1016/j.atherosclerosis.2008.07.010.

[32] M. K. Shea et al., "Vitamin K status, cardiovascular disease, and all-cause mortality: A participant-level meta-analysis of 3 US cohorts," Am. J. Clin. Nutr., vol. 111, no. 6, pp. 1170-1177, 2020, DOI:10.1093/ajcn/nqaa082.

[33] S. R. Zwakenberg et al., "The effect of menaquinone-7 supplementation on vascular calcification in patients with diabetes: A randomized, double-blind, placebo-controlled trial," Am. J. Clin. Nutr., vol. 110, no. 4, pp. 883-890, 2019, DOI:10.1093/ajcn/nqz147.

[34] V. M. Brandenburg et al., "Slower progress of aortic valve calcification with Vitamin K supplementation: Results from a prospective interventional proof-of-concept study," Circulation, vol. 135, no. 21, pp. 2081-2083, 2017, DOI:10.1161/CIRCULATIONAHA.116.027011.

[35] F. E. C. M. Peeters et al., "Bicuspid aortic valve stenosis and the effect of vitamin K2 on calcification using 18F-sodium fluoride positron emission tomography/magnetic resonance: The BASIK2 rationale and trial design," Nutrients, vol. 10, no. 4, 2018, DOI:10.3390/nu10040386.

[36] A. J. Van Ballegooijen, S. Pilz, A. Tomaschitz, M. R. Grübler, and N. Verheyen, "The Synergistic Interplay between Vitamins D and K for Bone and Cardiovascular Health: A Narrative Review," Int. J. Endocrinol., vol. 2017, 2017, DOI:10.1155/2017/7454376.